(Photograph by Susan Hampton)

Margaret Somerville lectures in the adult education program at the University of New England where she enjoys the local area and travels regularly to the nearby coast for research work. She has four children and has lived in many different Australian landscapes from inner city Sydney to the remote Tanamai desert. Her passion for landscape has been expressed over many years in several major collaborative research projects with Aboriginal women including *Ingelba and the Five Black Matriarchs* with Patsy Cohen, and *The Sun Dancin'* with four Aboriginal women from Coonabarabran. She is currently involved with archaeologists from the University of New England in an industry collaborative research project with Yarrawarra Aboriginal Corporation where she is collecting oral stories about place to produce an information database for their ecotourist industry.

Other books by Margaret Somerville

Ingelba and the Five Black Matriarchs
(with Patsy Cohen)

The Sun Dancin': People and Place in Coonabarabran
(with Marie Dundas, May Mead, Janet Robinson and Maureen Sulter)

BODY/LANDSCAPE JOURNALS

Margaret Somerville

Melbourne

Spinifex Press Pty Ltd
504 Queensberry Street
North Melbourne, Vic. 3051
women@spinifexpress.com.au
http://www.spinifexpress.com.au/~women

Copyright © Margaret Somerville
The individual stories contained in the work remain the property of the tellers of the stories, and the copyright and moral right are owned by the aforesaid individuals. Permission to reproduce these stories must be sought from the Publisher.

Copyright on layout, Spinifex Press, 1999
Copyright on all photographs remains with the creators

First published by Spinifex Press, 1999

All rights reserved. Without limiting the right under copyright reserved above, no part of this publication may be reproduced, stored in or introduced into a retrieval system, or transmitted, in any form or by any means (electronic, mechanical, photocopying, recording or otherwise), without prior written permission of both the copyright owner and the above Publisher of the book.

Copying for educational purposes
Where copies of part or the whole of the book are made under part VB of the Copyright Act, the law requires that prescribed procedures be followed. For information, contact the Copyright Agency Limited.

Edited by Gillian Fulcher
Typeset in 11/14 Palatino by Palmer Higgs Pty Ltd
Printed by Australian Print Group

National Library of Australia Cataloguing-in-Publication data:

Somerville, Margaret.
Body/landscape journals.

Bibliography.
Includes index.
ISBN 1 875559 87 6.

1. Women – Australia – Social conditions. 2. Social groups.
3. Landscape – Australia. 4. Aborigines, Australian – Women – Social life and customs. 5. Body, Human. 6. Oral history. I. Title.

305.89915

Contents

Introduction
    Body/Landscape Journals   2

Performance I
    Pine Gap Women's Peace Camp   18

Performance II
    Two Women Dreaming   44

Performance III
    Emily and the Queen   76

Performance IV
    The Sun Dancin'   108

Performance V
    La mer/la mère: and the cassowary women of Mission Beach   146

Performance VI
    Houses: and the performance of home   180

Notes towards a practice of love   218

References Cited   227

Index   234

Like a shadow on a landscape
Like a footprint in the sand
Like smoke across a mirror
This woman walks
This woman walks with a listening heart
This woman walks in the land.
(Unpublished song. Kerith Power 1996)

Acknowledgements

There are many Aboriginal women who have told me their stories whom I would like to thank including Nganyinytja, Emily O'Connor, Kathy Hinton, Maureen Sulter, Marie Dundas, May Mead, Janet Robinson. There have also been many women involved in the production of this writing but I would especially like to thank Bronwyn Davies, Kristine Plowman, Cathy Carmont, Laura Hartley, Sylvia Martin, Lizzie Mulder, Barbara Holloway, Linda Barwick, Susan Hampton and Kerith Power.

The publication of *Body/Landscape Journals* has been assisted by the David Phillips Scholarship awarded by the University of New England.

Introduction

Body/Landscape Journals

I begin with my body as you do with yours. It contains everything I am. I look downwards over my knees to my feet and then further downwards at my breasts and then across at my collarbone from shoulder to shoulder. This is not bleak, this is everything I am (Couani and Brooks 1983: 63).

Late summer 1999
Great phalaris grass seed heads, high as my shoulders, puff clouds of finest gold dust as I brush through their narrow passage walking out the back of town. The sun has stolen its colour, papery pale leaves and stalks rustle as breeze ripples gold waves across skin of paddock. Sweet smell of dry grass broken by pungent eucalypt leaves tattered by Christmas beetles iridescent on the ground. Down the hill into the green of sticky paspalum catching legs in summer shorts. A seedling plum drops rich fermenting maroon-purple fruit hollowed out by little birds and a hawthorn glows red with tiny fruit not yet ripe. Climbing up the other side, it's a land of grasses, wallaby grass, wild oats, wire grass, kangaroo grass, mixed with heads of plantain, dock, thistles and paper daisies. The sky is big and the mind opens out into layers of bleached paddocks broken by dark ruffles of eucalypts along fault lines and ridges. I turn and face the other way to see rim of indigo hills in play of light and shade as Mt Duval sleeps, inky blue.

Wind blows freely now, rushing loud through treetops, drumming on ears. A pale grey kite spreads its wings and plays in the currents. Black cows mosey up to the fence to see why I stand still. Down the other side seedling apples, green

Kungali Napaltjarri and friend, Illpilli, 1976
(Margaret Somerville)

and red, shine in leafy hollow and as I reach in to get one, a tiny wren ruffles and cheeps to tell me its nest is there. Bite into juice and tang of white flesh. Tumble into massed shoulder-high umbels of white grandmother's lace and satin sheen of paper daisies. A landscape bride. Posy of burnished copper and gold grasses, lace and satin white; a golden rosella feather, a pale green papery mantis capsule, and iridescent green and purple beetle wings to put on the desk beside me.

What is it about this place that I love? It takes me out of town to where the landscape unfolds from bitumen grid into hills and valleys. It is decidedly tatty, no picturesque white-trunked gums here, but a hybrid mixture of straggly black-butt, wild fruit trees, native grasses and weeds. I can know it by moving through it, heavy mud underfoot when raining—and puff of lungs uphill. There is feral food to gather—anyone can walk here and gather food—and they do. I love the distances that open out my mind, and the detail—of papery lantern mantis case and ants building up their nest—it must be going to rain today. I inhabit it through its different seasons, weathers, times of day and night, over months and years. There are other more spectacular walks in gorges and by inland rivers, but this very ordinary walk, the one I do daily, this imperfect, infinitely colonised space is 'home'. A space where I can belong, a pathway of desire (Neville 1999). But is it? This hybrid place, the in-between, represented here by the marginal, the not-owned, the publicly accessible spaces where anybody can tell their stories, is the focus of this question of belonging.

This place exists here in my performance of it. In telling the story of place it comes into being as a particular landscape evoked by a particular body, just as I come into being through that performance. How do I represent myself and the landscape? A farmer (Andrews 1998) tells a different story. He shows me the grasses in his paddocks when I visit his property. How the native grasses—barbed wire grass, native millet, lovegrass and kangaroo grass—are coming back through new methods of farming; the way his cattle won't get fat until the grass has hayed off in late summer; and how the long dry grass protects the soft green grass underneath from the toll of the summer sun. Or the big sweep of the paddock where the cows have eaten off the maroon seed head of silky brown top and the colour changes fom maroon to green so he knows when to move them on.

What stories does mine make space for and which ones does it displace? There is still an overarching sense that all the landscape is marked by Aboriginal stories and there has been no resolution to the questions whose land? and whose story can be told? Around the corner from the landscape of my walk there is a hill that Maisie Kelly (1998) tells me is one of the Anaiwan brothers was killed during a fight. As far as I know that story isn't written down. I have representational privilege—a computer, and a job that sometimes gives me time to write, space to go for a walk, access to publishers and I have been educated into the privileges of the world of writing. Does my story write out another story? Does it make room for multiple stories? Can your story be written in here? Is it a postcolonial space?

In his analysis of the technologies of colonial expansion, Benedict Anderson (1983) refers to the importance of 'the census, the map and the museum' as effective tools of colonisation. 'The map and the chart presented the colonial powers with new means of taking possession of land through the renaming of land and sea scapes' (McConaghy 1999: 34). Liz Ferrier (1990) suggests that colonisation is primarily a spatial conquest and postcolonial transformations require new ways of understanding and representing ourselves in space. She suggests that postcolonial transformations involve, in part, inscribing the body in space.

> The transformed cartographic and architectural models ... inscribe the body in space and suggest the corporeality of power/ knowledge, the possibilities of body knowledge, rather than perpetuating the Cartesian split between mind and body (Ferrier 1990: 182).

These and other such questions of belonging have occupied my thinking for a long time. There are two stories which help explain where this passion for landscape comes from.

The first was told sitting around the table in my sister's kitchen in Sydney. The story was in my bones but I had never heard it before and it explained a lot for me about the work I was already deeply engaged in. I had never heard about Wee Davy until then.

The story goes like this. My Nanna, a ramrod-straight steel grey Scotswoman when I knew her, came to Australia from Scotland as a young servant girl and married my grandfather, a Scottish carpenter. Although very poor she still wanted to return to Scotland to give birth to her first baby, but when completing the paperwork for travel she discovered that Papa had been married before and there was a child of that marriage in Scotland. Nanna visited that child on the Firth of Forth and was persuaded by relatives to bring him back to Australia. When she arrived with the boy, Papa was furious and there was 'trouble'. Wee Davy was placed in the Barnados Homes, never to be heard of again. He was only three.

This story is a kind of promise of connection that is lost. It represents all the loss and, for me, a generational cycle of erasure and repression of connection to place. In Australia, there is a double displacement: no Celtic indigenous to return to and, as a third generation migrant, I still bear the burden of guilt for the loss of indigenous here. So there is no choice, I have to flesh out a connection to place here because it is the only place I can; I have to make sense of that.

In the beginning of *Dingo Makes Us Human*, where Deborah Bird Rose (1992) talks about *remembrance*, she writes:

> The conquest of Australia was born in the oppression of the poor and dispossessed in England, Scotland, Wales and Ireland. Those in power assigned the cause of social problems to those who suffered most, and sought to alleviate problems by getting rid of people: transporting them to the Antipodes (Rose 1992: 1).

And Anne Noonan (1996), independent scholar, human rights activist and Jungian analyst, says that the Australian psyche is characterised by the archetype of abandonment, ejected from a motherland to which we can never return.

The other story, which is directly opposed to the Wee Davy story in its richness and presence, is from my time in the desert before I began any formal work with landscape. I had left Sydney as a young mother with three small children to accompany my husband to a position in a school on a government settlement at Papunya, two hundred odd kilometres west of Alice Springs. Because I did not have a government job and had access to a four-wheel drive vehicle I was much in demand to drive a group of older Pintubi women to their dancing grounds. There we spent the day dancing, singing—doing ceremonies.

~

I remember our last time together. We are sitting in a makeshift camp at the place where the women dance; a few sheets of iron, half a dozen dogs, six women, clothes and dilly bags hanging off the corners of the humpy. Sitting on the red earth around a fire with the old women I dance with, and trying to tell them I am leaving. We have driven out to their country and danced together often but this will be the last time. I know only a little Pintubi and they even less English, and I say that I am leaving because my husband has decided to leave his job. I have thought of staying on alone but am daunted by the fact that I am expecting my fourth child and have no means of support. They have never commented on my pregnant belly before, except maybe to pat it when putting on kangaroo fat for dancing. But this time, when I tell them I'm leaving, Kungali Napaltjarri says '*Palya! ninnaya Nungarrayi*, it's all right, just sit down here with us, he is going to get another wife.' What can I say? I know I will never see them again, nor can I write to them. They don't read or write. Shortly afterwards we left the desert.

The image of the women dancing grew with me and asked many questions. The women were powerful, dignified and in command in their place in the landscape. It was clear, as I read later in *Daughters of the Dreaming* (Bell 1983), that these women had a strong power base in land. When we returned to the east and I began to work with Aboriginal women in Armidale there seemed to be quite a different story. Most of the sites were taboo for women and yet they still seemed to be strong in their culture. What were their stories of place? How could I find out the stories of the landscape of the New England Tablelands? Did women have stories here? Was there a possibility of belonging through and with Aboriginal women's stories of place?

At this time Patsy Cohen approached me to help her research *Ingelba and the Five Black Matriarchs* (1990). I became the learner as the bare bones of the idea were fleshed out in the increasing richness and depth of our conversations:

Margaret: [When] I said what made you interested in finding out more about Ingelba and you said your identity. How do you explain that—your identity?
Patsy: Well I didn't know who I was. I knew I had black blood in me but I didn't know whether it was—I knew it wasn't Chinese—I thought to meself I wasn't Aboriginal 'cause Joey Goolagong was jet black and I didn't compare meself with 'im. Oh no, 'e's an Aboriginal so I can't be that and I did see Maoris and Hawaiian girls and I put meself down as one of them I suppose (laughing) 'cause I wouldn't identify meself with Joey Goolagong 'cause 'e was jet black and we were the only coloured kids in Bidura ... then and all the others were fair-haired—white kids.
Margaret: But why locate your identity at Ingelba?
Patsy: Well that's where the shock hit me—Woolbrook. Woolbrook Station was the biggest shock but then they lived in better homes at Woolbrook. They weren't the homes I was used to but at least they were [houses]—when I got out to Ingelba and got into the shacks out there and carried the wood

and things I've never done in my life before (Cohen and Somerville 1990: xi).

For Patsy, place had a central meaning in the construction of her identity. We entered into a process of learning and changing together in the complex crossing overs that followed. I learned to listen and hear the richness of stories in the hybrid landscape of Ingelba with its sheep, Patsy's grandmother's thyme, artefacts of its 1900 settlement, remains of hearths and fruit trees and the winding McDonald River always good for a billy of tea. Patsy had someone to listen to her and support her in a process that was emotionally important for her. She said that she had never spoken about these things before and that it had taken a load off her mind, that it would make a difference to her grown-up kids whom she had failed because of her anger, and her long and bitter struggle with her identity. In coming back to Ingelba and recording the stories, making *Ingelba and the Five Black Matriarchs*, she knew who she was. I think it was probably important to Patsy that I was a white woman hearing her stories, and I in turn felt that I had been born in this landscape, its Aboriginal stories were inside me. It was research through conversation across the space of our similarities and differences.

We tried to represent this relationship by expressing Patsy's and the other oral stories exactly as they were recorded so the text appeared with multiple voices. Some difficulties with my voice surfaced when reading with Patsy from *Ingelba* at the Perth Fringe Writers' Festival. I became painfully aware of the separation that the academic voice entailed and decided that it was not where I wanted to be located in the landscape of these stories. I was still committed to the idea of multiple voices and saw the developing process of representation of the oral as complementary to the growing body of literature by Aboriginal writers. There were many Aboriginal voices with oral stories that could be spoken into text in collaboration or research partnership. I began to see the academic voice as only one of the many possible voices that I could assume—and I wanted to strive for the inclusion of different 'I's' in the text.

The Sun Dancin' (1994) was negotiated with four Aboriginal women who approached me to write the story of Burrabeedee when we met at an archaeology dig. The focus of the story was another hybrid space, fifteen kilometres out of Coonabarabran, western NSW, where a large number of Aboriginal people had gathered to live at the turn of the century. It is now the very special site of Burrabeedee cemetery where we sat in the shade of a native cypress to begin our story:

Margaret: Tell me how to put it in the beginning?
Marie: You put it how you wanta.
Maureen: Say 'Long long ago in the dreamtime,' eh (laughing).
Marie: We don't wanta tell you how to do it Margaret. We want you to do it and bring it back to us.

Maureen's ironic reference to the way (traditional) Aboriginal stories are told reveals some of the complexities of the relationship between their oral stories and the written discourses that constitute Aboriginal people and impinged on our storytelling. They are aware that they are regarded as people without a culture because they have no 'dreamtime', having been robbed by the white colonial past which I represented. On the other hand they wanted to give me absolute authority in telling their story. They wanted me to go away and construct their story from the tapes we had made and bring it back to them for checking. They saw this authority as residing in me as writer, 'the pencil' as opposed to 'the mouth':

Margaret: So where do you want your story to start off Marie, cause we've got lots of good stories about your life?
Marie: We don't want to tell you how to do it Margaret, we want you to bring it back to us.
Maureen: Then you can have a look at it then mate (to Marie).

Marie: That's right, it's you who's making it for us. We don't want to say, 'Oh Margaret, you put that chapter over there, and that one up there.' We want you to do it.
Margaret: Yeah yeah, all right.
Marie: You're the person with the pencil; we're only the mouth, and the mouthpiece has gotta take notice of the one who's putting it in. We don't care where you put it.

Marie coined the metaphor of *the pencil and the mouth* to describe the interdependent relationship we developed over the representation of their oral stories as written text. The relationship between us was a constant theme in our conversations and negotiations about the book:

Maureen: You're the one who's the historian, see. I didn't know, I thought you was a—like Wendy, what was that?
Margaret: Archaeologist.
Maureen: All the time I thought you was an archaeologist.
Marie: I thought you was a digger too (laughs).
Maureen: She's diggin' all the time. I thought she was an archaeologist, and I found out she was an oral historian.

The women gave me multiple selves, the different I's I wanted in the text: the pencil as opposed to the mouth, archaeologist, historian, oral historian, and so on, but the new question was how to write a bodily presence?

The body/landscape journal itself grew out of a crisis of the body. In the initial phase of *The Sun Dancin'*, I was suffering from exhaustion and trying to write a paper about *relationship to place*. I knew there was another dimension that I wanted to write about but couldn't articulate. I could just hear echoes of Bill's story about collecting the grasses that blow up against the fences in Autumn before the cold weather comes:

I've seen 'em too go and get that—
you might see a lot of grass up against a tree
a tree or a fence
very soft grass

and it would blow like the wind you know.
They'd go and get this
and they'd stuff it into bags
and they'd make bed ticks out of them.
They used this a lot, dry grass that catches up on the netting fences,
they used that as a bed tick
it was just as soft as a bed to sleep on
it's very warm in winter
because it warms up and keeps the heat.
This is what a lot of them done you know—
this is about what I know.

(Bill Lovelock 1987)

I sensed the body and body/place connection always already there in the stories but didn't know how to do it for me. I thought and thought about the problem intellectually, and the more I thought the more distressed I became. I felt weak and exhausted, my heart pounded, even to walk up the stairs was an effort. I had strange and frightening dreams about a fragmented body—one in which my body was sliced into fine layers of flesh which were cooked and spread with vegemite. I was trying to think through complex theoretical problems and could think no longer. I had fallen into the abyss of Western dualistic thinking predicated on separation rather than connection.

I tried resting and yoga to get back into my body but with little change until a friend asked me to walk in the gorges. How could I walk into the gorge when I couldn't walk from the bedroom to the kitchen? We walked down and down and down the winding path to the black water-coursed rocks of that hard place where we rested, and then abandoned made paths to slip and slither our way through fern gullies and shale slopes back to the top. The panting, heart-lurching, slithering hardness of the walk forced me back into my body. Richard stopped from time to time to make philosophical notes in a tiny pocket notebook while I concentrated on survival. From this walk I understood that I needed to repair a profound mind/body split and that in this lay the extra dimension of body/place that I was unable to articulate. The absence was the body, 'the most alien of dragons lurking on the edge of our known world':

> Even circular maps of the world have blank spots which we prefer to dismiss as the haunts of monsters and other classifiable entities rather than beings with whom we must form a relationship or conduct a dialogue. Bodies themselves, the traditional bedrock of material reality, may be the most alien of dragons lurking on the edges of our known world (Gunew 1991: frontispiece to Diprose and Ferrell).

I began a series of massages with Cathy Carmont who provided a space of trust where I could risk madness, speaking images that surfaced from tissue, organs, joints and bone, without the need to make sense. There was no system, no order, no expectation; just pleasure in the touch of skin, release from pain and tension and permission-giving in whatever verbalising (or silences) I chose. Cathy recorded our massage talk for her own research on massage (1996a):

Margaret: It makes me think about what you said before—about moving mountains by going underneath, or tunnelling through or lying down or something rather than actually having to shift whole mountains with your shoulder.
Cathy: What you said before? Being in a landscape.
Margaret: Yeah, cos I didn't think of that connection. But that's like I feel now—that you can lie down real low beside the ledge of the mountain and shift it.
Cathy: Yeah.
Margaret: Instead of before—I had this image of these big mountains that I had to push the whole mountain.
Cathy: Yeah.
Margaret: Head on my shoulders.
Cathy: From your upright position.
Margaret: Yeah.
Cathy: Which is your strength.

Margaret: But now what I'm learning is that you can lie down real flat and low under the ledge and move the mountain.

This technique of expressing images which occurred in the in-between space made by the movement of body surfaces—Cathy's hands over my skin—allowed a play with new ways of representation. I started to keep a journal by writing down some of these images immediately after the massage:

A tight place in my tummy, a hard, black rock about fist-size, round and flat. As Cathy massages, the rock is washed with water which wells up into throat and eyes. A deep sadness. It belongs to the women from Coona undoing knots of connection made when two souls bump into each other and then always having to leave.

I want to go to the splashing place where the water splashes over hard smooth rocks. Rock inside and rock outside; place in body and body in place.

The journal writing was reluctant and spasmodic because I had to remember the images during the massage, and these images were often not articulated into words. It meant bringing myself out of a deep trance-like consciousness to cross over the bridge between semiotic and symbolic, to give the images words, and then to remember them. I decided it might be possible to reflect on the experience of my body in other places:

Picnic with kids; little rocks in the shade inviting but they want to play on big flat rock in the sun. Warm and sleepy, I lie in shallow dip on lichen-dimpled surface of big flat rock, face into brightness. Circles of green light form and reform behind my eyes. My skin becomes flecked with grey and light apple green lichen-lace and I am a lizard, grey scales and flecks, lying on a rock in the sun. I become so flat I am the rock, body blends into its surface, tufts of soft green moss around my edges and voices of children playing over me. I am the surface of the earth and they are playing on my edges.

A new pleasure came with this writing. One day watching the waves scattering patterns of shells and stones across the sand and then shifting them with the next wave I felt I could write with this sort of ease. I learned to play with language as Fallon does:

> lingalonga over lingua
> you leave me reader working on the body of my new
> lover Trixi oh sorry what was it again Lexi yes Lexi Con
> holding her spine in the palm of my hand
> ah and ohyes the body of language
> (Fallon 1989: 32)

The journal writing grew through the period of my work with the Coonabarabran women, often getting its focus from my experience with them and in their landscapes. It tended to evolve side by side with recording and transcribing their stories and thinking about how to represent all this in text. I was walking in Coonabarabran and wondering about my presence when an image came to me:

Who am I in relation to these women and their stories? Is it four women or five? My consciousness is moving in and out from thinking about the book to the texture of gravel under my feet, clouds flickering across the early morning sky and the cool lightness of air on my face, when I notice another presence. It is elongated and slightly distorted in the characteristic way of a Drysdale bush woman, and it walks with me. The sun is rising in the Eastern sky and has cast my shadow on the ground in front of me in what seems like a powerful image of my place in the landscape of these stories. There has always been a prior question for me of my place in this Australian landscape.

The shadow represents fairly closely the extent of my bodily presence in the text of *The Sun Dancin'* (Somerville et al. 1994) and has become a pivotal image for me. The shadow treads lightly on the land of these stories, you can still see the fine details of gravel and grasses through its shape. Although the integration of my own journals with the stories in *The Sun Dancin'* was slight, its importance for me was to make myself physically and materially present in the landscape of these Aboriginal stories. I argued both for myself and potential readers: if there is no chance of having an embodied presence in the landscape of these stories, then how are we to develop relations of empathy with the tellers or with the landscape? How can we understand the magnitude of dispossession and the possibility of new stories? Or are we to be forever aliens in this landscape of Australia? In Marie's words, *we've gotta make it good for ourselves to go forward*.

When I began what became the *Body/Landscape Journals* the purpose was to revisit the people, places and stories of my work to explore the notion of bodily presence. My memories were like clusters of dots from a dot painting in the landscape of my mind, each of the places a shimmer of dots that could be entered at will to become the whole world. I approach theory like a bower bird, choosing fragments by colour, taking them into my bower and leaving the rest. Some special pieces, like the prized blue plastic helicopter propeller I recently saw a bower bird carry with him from bower to bower, go with me from place to place and link the stories together.

These linking theories are the interrelated ideas of performance and liminal space. I begin with Pine Gap Women's Peace Camp because it seems to contain all the threads of later work. Although it had been transformative in making women visible in the landscape it had not been articulated as an experience in itself. To understand the process of women making themselves visible in the landscape, I use Victor Turner's work on performance from which emerges his idea of liminal space, as Schechner (1986: 8) describes it: 'performance is the art that is open, unfinished, decentred, liminal'. Turner draws his ideas about drama primarily from African rituals such as initiation where the purpose of the drama is a bodily enactment of the transformation from one state of being to another. The performance creates an in-between or liminal space, a space of possibility where the individual is in transition, moving from one state to another.

He extends this understanding to include what he calls social drama—protest, festivals, carnival—in which the participants are challenging the established order. In the liminal space of social drama the established order is turned on its head so that new possibilities can emerge.

In the chapters that follow, each set of experiences of women in the landscape is conceived of as a performance. Alongside this idea of performance runs the parallel idea of the liminal or the in-between. This idea has been developed in a number of different ways by widely diverse theorists, the most notable being Paul Carter for his work on the space in-between the indigenous inhabitants and the new settlers.

> Their value lies ... in opening up a space between and around them, a dynamic space that ... kept all the views open; that preserved the intervals of difference (Carter 1992a: 179).

In Pine Gap it is Turner's idea of a space of becoming performed in social drama; in Two Women Dreaming it is the space between self and other theorised by Elspeth Probyn; in Emily it is Paul Carter's choreography of the dance of differences; in Sun Dancin' it is the space of oral storytelling, between the pencil and the mouth; in La Mer it is Kristeva/Irigaray's space between body and language, and in Houses it is Ferrier's space of postcolonial transformation where the postcolonial house folds the outside in.

It is in the accretion and intertextual spaces of these various diverse approaches to the liminal that the meaning of the whole lies and gathers together into the final question. What in theory is able to respond to the power of Aboriginal (and white) women's stories of place? How can we formulate a practice of love not separation? The result is Notes towards ... fragments of theories and stories to be finished by us all.

Performance I

Pine Gap Women's Peace Camp

Landmarks[1]
Pine Gap, *1983*

We are camped on a narrow strip of dirt between a straight bitumen road and a new five-strand barbed-wire fence. The road leads to metal gates and watchtower and beyond, the white radomes that we never see. In our mind's eye we hold an image of the sleeping red rock at the centre but we don't see that either. On the other side of the fence is no-man's-land, a more gentle desert landscape of grey green mulgas, spinifex and scatterings of low growing shrubs and flowers, red dirt with creeping strands of paddymelon vine with green- and yellow-striped fruit. Somewhere else, the Aboriginal women who have joined us in protest, are camped. For three weeks this roadside strip of freshly bulldozed red earth is home. We live on the ground, the luminescent blue fabric of our tent the only membrane between us and heat, glare, noise, swooping night helicopters, hundreds of other women—and red dirt which clings in sweat between our toes, seeps into clothes and changes the colour of our skin. We are here doing our daily lives—cooking, washing, sleeping, shitting—and protesting against the presence of the American military surveillance facility in the centre of Australia. Ten years later someone says, the white radomes are still there, what did the women achieve?

---

1 I thank Gillian Fulcher, my editor, for this idea and for what it says about the book's theorising.

*Women carrying gates of Pine Gap Defence facility, 1983
(Phillipa Nightingale)*

In this performance I explore the experience of Pine Gap through photographs, writings, memories. What meanings can we give it now? Using the methods of the body/landscape journal, the process of memory work unfolds, becomes circular and layered as memories are recovered and theoretical questions posed. Turner's work on social drama (1987) gives me the concept of liminal space around which to develop a framework of meaning for Pine Gap. As a liminal space I represent the experience of Pine Gap as the in-between, a space of transformation, where you are no longer in the state you were, but haven't yet arrived at a new state. The process of writing itself evolves from this liminal space, using daydreams and night dreams, stream of consciousness sequences and body feelings, to begin a process of representing landscape differently and to tease out the threads that stretched into later stories.

~

> The social drama is an eruption from the level surface of ongoing social life, with its interactions, transactions, reciprocities, its customs for making regular orderly sequences of behaviour. It is propelled by passions, compelled by volitions, overmastering at times any rational considerations (Turner 1986b: 90).

In November 1983, the *Sydney Morning Herald* reported that 'seven hundred women travelled from all over Australia to the Pine Gap Women's Peace Camp to protest against the presence of the Joint American and Australian Defence Facility at Pine Gap, fifteen kilometres out of Alice Springs'. I was there, we were represented in the media, but most of us have been strangely silent, unable or unwilling to say what the experience of Pine Gap meant to us. What can I say of this experience and how can I represent it to you?

Haug (1987: 71) describes the process of precise remembering as recovering a key image and then recalling every detail—'or a smell, perhaps, a colour, sounds or music'. I will begin with some memory work.

I remember
making new tracks
women in four buses shining silver
snail trails winding
from coastal margins
to a circle in the centre,
the *ngurra*[2] of Aboriginal paintings.
It could be Uluṟu
the great sleeping rock of the red desert country
or is it Pine Gap Defense Installation
with round white radomes
a centre of silence.
Or maybe the circle is an egg
passed from policemen to women protesters
or a peppercorn wreath,
a circle to remember our forgetting
laid at the gates
where hundreds of circles of hats and parasols
confused straight lines of roads and thought.

~

I am a young mother with four children and have never left them
heat dust fear noise sleeplessness
I am one among hundreds of women camped together on a strip
   of red dirt
beside the road
talking movement confusion action
I am fragmenting with images, my very cells change
excitement passion sadness closeness
I am Karen Silkwood
I remember singing

11th May

I talk with Mary about Pine Gap and words pour out, but when I write it is so slow. Written words are overlaid with other meanings, ecofeminism, mothers as nurturers, nurturing the land, too simple to account for the multiplicity of Pine Gap. No story at all is better than an easy

---

2 In Western Desert dot paintings the circle represents *ngurra*—camp, place, campfire, waterhole, womb.

oversimplification. It is more like Carnivale, antistructure, a temporary abandonment of meaning—like Goldwasser's ideas, as Turner describes them:

> Antistructure is represented here by Carnaval, and is defined as a transitional phase in which differences of (pre-Carnaval) status are annulled with the aim of creating among the participants a relationship of communitas.[3]
>
> 'To make a Carnaval' is equivalent 'to making chaos' where everything is confused and no-one knows where anything is (Turner 1986b: 132).

It is like the photograph, I say to Mary, of a child looking through the gates which have been decorated with ribbons and a teddy bear and a sign that says 'Women are Wild about Living'. It is about paradoxes and contradictions, the complexity of individuals present and yet their participation in a common experience.

Fusco also talks of carnivale:

> Carnival was the space where, for a day, disorder was allowed. That is what is so interesting about carnival: there were so many different eruptions ... Those different tensions were all placed in that space (Fusco 1990: 18).

What are the different tensions, in this space of carnivale?

I open a worn manilla envelope with forty photos and a few scrappy pages of writing, my entire memorabilia, and show Mary the photographs. There are wild women there, women like I'd never seen before, and women remembering their children, multi-faceted images existing simultaneously without the need for one to subsume the other. There are many other signs in the photographs—slogans on the gates, symbols of Karen Silkwood, the texts of the banners, and the subversion of form as well as meaning. To say something in a series of calico banners with cloth and thread and paint is quite different from saying it in words. Looking at them now,

---

3 Turner's notion of *communitas* stresses a particularly communal rather than individual suspension of meaning.

I remember many *small stories*, the story of the banners, the peppercorn wreath, the human peace symbol, women and police exchanging words, emotions and an egg. There are longer stories which I may want to tell—the dismantling of the gates, the night our tent was carried away in a wirli wirli, the story of Karen Silkwood, the Arrente and Pitjantjatjara women dancing.

But there are no photographs of the Aboriginal women. What does this absence mean?

I remember dancing in the desert, how I felt so much part of the performance that it was impossible to take photographs. But on the last day I decided to take the risk of separating myself out from the dancing just enough to record some of it. Although the women felt fine about it, I felt distanced and sad, but was leaving anyway and it would be my only record. It was almost as if the camera had to take the photographs itself, I had to be so quick and surreptitious. The photographs are shonky, taken from strange angles, depending on where I am sitting and show all the peripheral paraphernalia—like eggs in a blue cardboard egg carton that I brought for lunch, a big white plastic container of Saxa salt, the inevitable billy and the red, red earth, as well as the dancing women.

I look at these photographs now and remember the feel of the day. The heat and red dirt, smell of kangaroo fat on warm skin, high-pitched, rhythmical singing and beat of hand against thigh, thud and scrape of dancing; laughter as we fall about over our mistakes. Touching and leaning closeness, hunger pains quelled by too much sweet black tea, and after the dancing, tearing into strips of green kangaroo meat secreted under the seat of the truck. Here the women are giants in their own country. Later, back at the settlement, there are squalid tin humpies, and me cooking dinner in a concrete block house with only the dusty feel of ochre patterns on breasts as a reminder of the day.

It was these same feelings that were called up by the Arrente and Pitjantjatjara women dancing at Roe Creek, our sandy riverbed base camp, before we actually went to the gates. The women dance at dusk by the light of flickering fires and there is the same gesture of warm skin and touch; the beat of the dance seems to travel up through the ground into the body. They laugh and fall about as they negotiate their performance which is short, and then they are gone. It was said to be a Two-Women dreaming story which travels through Pitjantjatjara and Arrente country and includes the land where Pine Gap is now. As with my dancing with the women in the desert before, we have no means to interpret this dance, or the songline they are offering us, in terms of reading the landscape or sharing mutual concerns. And there are no photographs of the Arrente and Pitjantjatjara women at Pine Gap.

How do the photographs constitute this performance? Do they make apparent the particular workings of this social drama? I place them on the noticeboard to give some form to this data.

12th May, 2.30 am.

Moon-talk. I wake to a ferment of ideas. Is this what Laura means when she talks about the excesses of her premenstrual state? She tells me that women can have heightened states of awareness and wild proliferations of ideas, hormonally, just before menstruation. In my premenstrual migraine I usually feel clumsy and inept and want to shut the world out. But tonight it is full moon, I am bleeding, my belly aches with menstrual cramps and I feel like squatting and baying at the moon. I surrender to the flow of ideas—patterns rather than words. It is a wild connectedness where images merge and fire off each other and my mind explodes. What is this other kind of thinking that happens in the night?

> The moon is a sow
> and grunts in my throat
> Her great shining shines through me
> so the mud of my hollow gleams
> and breaks in silver bubbles

> She is a sow
> and I am a pig and a poet
>
> When she opens her white
> lips to devour me I bite back
> and laughter rocks the moon
>
> In the black of desire
> we rock and grunt, grunt and
> shine.
> (Levertov cited by Hall 1980)

~

Ishtar, the moon goddess, demands that we open our eyes to the *lumen naturale* that shines in poetry and dreams. She asks that we value the irrational, the instinctual, the uncommon sources of knowledge (Hall 1980: 17).

She represents the principle of transformation.

> Paying attention to her means admitting paradoxes of intention and action ... listening to inner voices or going to the oracle of one's own dreams when blinded by too constant light of day (Hall 1980: 14).

In the moonlight I have a vision of layers of maps superimposed on each other, as if drawn on tissue paper, so that all the layers are visible through the others. The first layer is the dancing places of Aboriginal women; it is Aboriginal women's mapping of place through story, song, dance, and site—linked by songlines to other sites across the landscape. The second layer is the superimposed one of white settlement at Pine Gap—road, fences, gates, gatehouse, and the symbolic white domes beyond. The final layer is made up of images of the women's peace camp, re-colonising the roadside with cloth, tents and banners, recasting the meaning of gates with flowers and leaves, remaking fence and road with new stories.

> ... palimpsest: ... an overwritten page, a script under which is shadowed another script, another text ... Thematically, morally, textually ... the erasure of the signs (mark, trace, index, imprint) of the 'mother' the text made marginal—by the signs of the 'father'—the text of dominance. But imperfect erasure. Can see both writings. Can see them as interactive (Blau du Plessis 1990: 86).

Old stories, new stories. I read Sarah's published Pine Gap story of childhood remembrance days, the eleventh of the eleventh, and am reminded, surprised, of The War. Our main march took place on Remembrance Day in a rewriting of the hegemonic spatial practices. I hadn't thought at any time of writing about the war. What is it that I want to forget? How does one remember, and how does one forget?

An image of a photograph comes to mind, of Mum and Dad just before the war, just after they were engaged.

In this picture my mother, in her early twenties, looks unusually tender, in love. I can remember the soft, pale green silk blouse and brown woollen skirt. She is arm-in-arm with my father, a handsome young man dressed in air force uniform, and they are striding down Pitt Street, Sydney. My mother had said as we looked at this photograph, 'I was engaged to Dad before he went to war; six years later he came home and we got married. *I was afraid that I no longer knew him.*'

The next year I was born, the first of four children, to a father with the memory of war in his bones.

In *Poppy*, Drusilla Modjeska describes a 'brooding' pattern of thought—circular, rather than linear, wandering rather than purposeful. I am now brooding on my mother, when, in a moment of perfect conjunction between inside and out, I reach for a sheet of scrap paper to continue this writing. There on the back of the paper is a poem by Maxine Kumin, who was born in 1926, the same year as my mother:

> She
> remembers especially a snapshot
> of herself in a checked gingham outfit.
> He is wearing his Navy dress whites.
> She remembers the illicit weekend
> in El Paso, twenty years before
> illicit weekends came out of the closet.
> Just before Hiroshima
> just before Nagasaki
> they nervously straddled the border
> he an ensign on a forged three-day pass

she a technical virgin from Boston.
What he remembers is vaster:
something about his whole future
compressed to a stolen weekend.
He was to be shipped out tomorrow
for the massive land intervention.
He was to have stormed Japan.
Then, merely thinking of dying
gave him a noble erection.
(Kumin cited by Heilbrun 1988: 62)

We lived The War all my life with our father. Uncle Tommy and Aunty Joan, Uncle Slim and Aunty Betty were not relatives but compatriots from the war. 'Three Squadron' was a family phrase. We heard veiled stories of Dad swinging from a chandelier in Naples and around our plain Presbyterian house were scattered trinkets from faraway worlds: perfume bottles from the middle east, embossed china tea cups from Turkey, exotic jewellery from Milan. On Anzac Day, Dad would disappear for the whole day and we would watch the march on television, trying to pick our father out from all the men in uniforms. Once we saw him and knew that he was there.

Today, my Mum is dead, and at 73 Dad still replays The War. The reunions seem to happen with greater frequency; he produces the *Three Squadron Newsletter* and has researched the history of Three Squadron. He attends sick beds and funerals, cares for the widows or widowers of his fast dying air force friends, and every place he visits is inscribed with The War. There are networks of them all over Australia, in every town I have lived in, and where my children now live, there is always someone to visit, something to organise, in memory of the war.

And I married a man whose father also fought in The War. The one photograph we had of Frank is of a handsome, full-lipped young man in soldier's uniform. Another war this time, a bitter and more bloody battle. Frank fought at Gallipoli in World War 1 and his only story about The War, which Ellen (his wife, my mother-in-law) told us, was about the trenches in France. He was fighting in the trenches with his best mate, and he watched as his mate stood up and got his head blown off. Frank didn't glorify the war, he had repeated nightmares right up until he died, a very old man. These are the stories that no one tells, the stories each man has to live with alone, without resolution.

Toni Morrison, writing from the margins in America—black, woman—imagines such a story from the inside:

> A young man of hardly twenty, his head full of nothing and his mouth recalling the taste of lipstick, Shadrack found himself in December, 1917, running with his comrades across a field in France ... He ran, bayonet fixed, deep in the great sweep of men flying across the field. Wincing at the pain in his foot, he turned his head a little to the right and saw the face of a soldier near him fly off. Before he could register shock, the rest of the soldier's head disappeared under the inverted soup bowl of his helmet. But stubbornly, taking no direction from the brain, the body of the headless soldier ran on, with energy and grace, ignoring altogether the drip and slide of brain tissue down its back (Morrison 1987: 7).

Frank tried to shut off part of his brain to forget, and brought his children up without emotion, in fear and violence, like an army major, to keep the pain at bay. But he never forgot. I always understood that, in a way, it wasn't his fault, he was just continuing a pattern, holding the pain at bay. I went to Pine Gap to protest about men's violence.

> ... for the problem of memory and forgetting, for those who would address it, lies in determining at which point one process—the process of forgetting—should end and another process—the recovery, the deep remembering—begins (Brooks 1991: 34).

13th May

I have now placed the Pine Gap photographs on my noticeboard and they allow me to go a-travelling. The photographs themselves move around as I group and regroup them trying to understand what they mean. What is their message? How do the photographs themselves constitute the experience of Pine Gap?

It is strange that the one memento I have is the visual image, the particular art form which is renowned for its distancing, separating seer and seen. I read works on photography and realise that the art photograph is taken as the norm, the universal. Most of my work has involved popular photographs—the Aboriginal women I have worked with all have them, there are photographs of every landscape performance, holiday snapshots, family albums. Popular photography is also, in my experience, most often used by women as a form of cultural production and a means of reading and transmitting their (alternative) stories. I decide that in reading the text of these photographs it is necessary to reflect on these genres of popular photography.

The popular photograph enables everyone to construct their own image in the landscape. I recall watching, and wondering, as Aboriginal students on excursions stop to take their photographs standing beside the inevitable signpost—'MT ISA'—with three women standing underneath. This most common form of popular landscape photography seems to be concerned with a confirmation of presence: *I was here*.

> I can never deny that *the thing has been there*. There is a superimposition here: of reality and of the past. And since this constraint exists only for Photography, we must consider it by reduction, as the very essence, the *noeme* of Photography (Barthes 1984: 76).

The meaning of this popular form makes it an eminently suitable form and genre to represent a performance of alternative spatial practices in the landscape. And, rather than distancing, it seems to be more concerned with establishing a connection between photographer and photographed, subject and object, seer and seen.

> the noeme 'That-has-been' was possible only on the day when a scientific circumstance (the discovery that silver halogens were

sensitive to light) made it possible to recover and print directly the luminous rays emitted by a variously lighted object. The photograph is literally an emanation from the referent. From a real body, which was there, proceed radiations which ultimately touch me, who am here; the duration of the transmission is insignificant; the photograph of the missing being, as Sontag says, will touch me like the delayed rays of a star. A sort of umbilical cord links the body of the photographed thing to my gaze: light, though impalpable, is here a carnal medium, a skin I share with anyone who has been photographed (Barthes 1984: 81).

I think about these things now and give you a description of the essence of each grouping the photographs have settled into, with an example from each one: my alternative to the family slide show.

(Meta)physics of presence

The first group of ten photographs is about this (meta)-physics of presence, that I have referred to above. Their major claim is that of the popular traveller's snapshot: *we were here*. In viewing them I recall the original experience, but they do not finally name for me an individual, or even a shared group experience, although that is one reading. The presence I see is a collective or community of women who decided to join together in the same place, at the same time, to make an alternative set of spatial practices visible in the landscape. The photographs show only women, a women's-only space but they also immediately suggest difference, the presence of a great diversity of women whose only commonality may be that they decided to be present at the same time and place, to protest together.

My eyes sweep over these surfaces and the images touch me as I touch them. There is a crowded pub verandah—stopping place on a long journey—and arrival at the base camp, spreading river gums and shade on white sand. There are meeting circles where women sit on the ground and plan their action. A red dirt road is covered with miles of cloth banners with figures of absent women from all over the country—painted, embroidered, appliquéd and knitted. Then there are real women swathed in patterned lace and soft scarves, textured hats and light cotton cloth against heat and sun, sitting on fallen mulga timber under feathery shade of the mulga trees, waiting to march. The milling of the women at the gates, where hundreds of circles of hats and coloured clothes inscribe a new image over straight lines of the road, fence and gates.

> Liminality can perhaps be described as a fructile chaos, a fertile nothingness, a storehouse of possibilities, not by any means a random assemblage but a striving after new forms and structure, a gestation process, a fetation of modes appropriate to and anticipating postliminal existence. It is what goes on in nature in the fertilized egg, in the chrysalis, and even more richly and complexly in their cultural homologues (Turner 1990: 12).

Transforming signs

The next group of fourteen photographs powerfully records the recasting of sign imagery, a new semiotics of place. Every woman seemed to carry a can of spray paint, pliers, bolt cutters, or a ready supply of cloth and ribbons in her handbag, for an immediate and spontaneous transformation of gates, fence, road and official signs.

Sometimes it is the simple defacement of public signs, such as the addition of the words *Pine Gap* to the Stop sign on the road, and in others it is the addition of many words and the erasure of others. Each action seems to generate its own sign dimension; the hanging of a peppercorn wreath on the fence to commemorate the loss of the land when the Australian government accepted 'a peppercorn rent' from the American government; the covering of the road and new gates with signs *Congrats on a hasty erection* on the night after their construction when the old gates were removed by the women. Graffiti appears as an ongoing comment on daily events and an impressive recasting of forms of power.

> One could think of a variety of discursive practices—the production of signs, names, maps, graffiti, representations or narratives connected with place—as practices which appropriate or transform space (Ferrier 1990: 36).

Cloth streamers and ribbons are used to reweave the fences in a similar transformation to that of the gates. Metaphors of weaving and spinning had become common in radical and transformative women's poetry of the early phase of the women's movement. Adrienne Rich evoked the web weaving of the spider to represent women's work in radical transformation. These new images are created in the landscape of Pine Gap by the continuing interaction of presence and performance. We are here resisting the given images, signs and storylines of this place, and we are creating new signs in this place (which we can carry away to other places?). Aboriginal women read the signs in the landscape that tell the story of their lives; at Pine Gap, women rewrite the signs in the landscape and read themselves back.

> [Social dramas are] reflective in the sense of showing ourselves to ourselves. They are also capable of being *reflexive*, arousing consciousness of ourselves as we see ourselves. As heroes in our own dramas, we are made self-aware, conscious of our consciousness. At once actor and audience, we may then come into the fullness of our human capability—and perhaps human desire to watch ourselves and enjoy knowing what we know (Meyerhoff cited by Turner 1982: 75).

Domestic images

There are seven photographs of our living space, and the familiar faces of the women I went with. In them I recognise the private emotions and growing friendships that sustained us. The local place groups are the most basic level of organisation, and thus the locus of personal change. In this community of women we deal with daily life in the camp, we laugh and cry together, and begin to tell new stories to each other.

The photographs themselves make apparent another aspect that was perhaps not so visible to us at the time. They show us sitting in the blue light of our K Mart gazebos, through the flyscreen walls other tents and living spaces beyond. Blue, green and orange light shines through the tent cloth; the landscape swathed in soft billowing cloth of tents with clothes strung on lines between them. There is a breaking down of boundaries between inside and out—the outside comes in and the inside out. Women, no longer confined to the invisibility of their private spaces, make visible the spatial practices of those private spaces in the landscape.

> But the performances characteristic of liminal phases and states often are more about the doffing of masks, the stripping of statuses, the renunciation of roles, the demolishing of structures, than their putting on and keeping on. Antistructures are performed too (Turner 1986b: 107).

The membrane that divides our inside from outside is not only permeable but fragile. As we lie in our makeshift camp one night, a gust of wind springs up and flurries through the tied-down tent flaps. Each puff and flap is followed by quiet stirrings as sleepy women tie and re-tie flaps and secure pegs, to keep our world in place. Towards dawn, as the big silver moon drops low in the sky, the wind eddies and gathers into a whirling funnel. Up fly knickers and hats, clothes and papers whirling in a spiral of red dust for a redistribution of earthly goods. The tent balloons and shrinks, and balloons and shrinks again, until finally, despite our trying to hold down our world, it puffs off and away. We laugh, exposed in sheet bags to the pale green morning light, and Lenore puts on her lipstick to greet the new day.

Narratives of action

There are eight photographs that tell the story of our actions. A series of four shows us dismantling the gates and there are four individual photographs of other actions.

The dismantling of the gates, two lines of outstretched arms are holding onto the gates, police officers on one side, women protesters on the other. Each group is leaning, straining towards the gates; it is clearly a photograph of opposition although there is no obvious sign of violence.

The gates are rocked back and forth, and as the rocking builds up momentum an angry energy is stirred, echoed in songs and chants. We sing and rock, rock forward and push, rock backward and pull, the police and women dancing in a rhythm of opposition. Then, quite unexpectedly, one woman calls out to pull instead of push. As the police push, we pull and the gates come loose into our arms. A few women carry the gates away and police link arms to form a human fence facing the rest of the protesters directly. No one challenges this barrier as the rhythm of our singing changes from anger to sadness: 'we are gentle angry women singing, crying for our lives.' The women in the police line have tears and do not appear in the front line after that. Later that day, a policeman offers a woman an egg, hardboiled, from his lunch as they talk about the action. The gates are the symbolic site of struggle and a lot is at stake. Even though we are still seven kilometres from the base, the gates represent a boundary, the threshold, a liminal space marking the inside from the outside.

In the second photograph a group of women carry the gates away on upstretched arms. Finally the gates, garlanded, and woven with wildflowers and native grasses, are placed at the entrance to our camp.

New gates are erected, and the garlanded gates removed as if the process can be undone. But images have already been transformed and a new story appears on the road the next day, a sign: *Pine Gap open by women for peace to the public, November, 1983*. The gates are the central site of contestation of meaning, and their meaning is transformed from legislative/exclusive to garlanded celebratory, the (always open) gates to our camp.

The social drama remains humankind's thorny problem, its undying worm, its Achilles heel—one can only use clichés for such an obvious and familiar pattern of sequentiality. At the same time it is our native way of manifesting ourselves to ourselves and, of declaring where power and meaning lie and how they are distributed (Turner 1982: 78).

The actions involve turning familiar stories on their head, re-telling old stories of resistance or making new stories. Changing the storylines involves complex strategies of deconstruction and reconstruction based on creative and collective action.

14th May

I am awake in the middle of the night again and listen for things unsaid. One final photograph haunts me with its insistent image. Why does this image return?

Of all the photographs this one is not especially distinguished as belonging to Pine Gap; it is the fortieth photograph, not grouped with the others. It shows a small child on the road alone, her smallness emphasised by the largeness of the ground surrounding her image. Originally I saw it as the only photograph that introduces the idea of nurturing; yet the child is alone, without a mother, small, vulnerable and beautiful. Even unsentimental friends respond to this photograph with loving sounds and tender gestures. Perhaps it is the image of a small child in the context of Pine Gap that makes it such a powerful picture.

But in the night this image is juxtaposed with another, very similar image of Jessica, my 'baby'. In this other photograph Jessica is about the same size, a fine fuzz of golden hair, alone against a large ground. She looks a bit waifish, fey, not quite of this world, a magic little person bringing out strong flights of maternalism, but equally strong admiration for the indomitable and independent spirit that lies in that little body. Jessica is fifteen now and nearly grown but still with an airiness, an evanescence, as if she is not solely of this world.

She wants to be a dancer. She lives with her father. I still haven't got any of my family photographs because I have stepped out of the story. I have no story yet, but I am making it. Can this photograph of a little girl alone in the middle of the road at Pine Gap be my photograph, my baby?

Yet this photograph is also of all my babies who have grown up and left home, of all our babies. It is not only that I have lost Jessica, but that it is inevitable (and necessary) that children grow and leave home. I grieve for all my photographs, for my children growing, for my own growing old. It is a photograph about longing. In it I take into my arms what is alive and transient, captured from a moment of time. For me this motherlove cannot be reduced to a rational explanation; it is the passion and the searing pain, both the agony and the ecstacy; ecstasy that moments had, agony in awareness of their passing. I juxtapose these feelings from the image of the small girl child to the context of Pine Gap; the opposition of the intent of global war with the experience of one woman as mother.

15th May

I try to retrieve more stories but they are full of holes so I watch with Jessie the video of Karen Silkwood. In *Silkwood*, Karen is an anti-hero, a young working class woman who is exposed to dangerous levels of radiation in the plutonium plant where she works. She eventually dies in a mysterious car crash while trying to expose the corruption of the big corporation. She doesn't set out to save the world, but is concerned with the immediate, dehumanising and painful treatment she and her friends have to put up with at work. At Pine Gap, Karen is the anti-hero of a new narrative with which we can identify. When eighty women break through the gates and are charged with trespass and taken to the Alice Springs watch-house on the anniversary of Silkwood's death, we all give our name as Karen Silkwood.

My daughter identifies with the young woman in this video and after it she asks about radiation poisoning and nuclear reactors. We talk about Chernobyl, the clouds of black radiation dust and the children who are still getting sick as a result of exposure to the radiation. She remembers the story of Hiroshima, which captured her attention at school. She had brought home a matchbox full of tiny paper cranes, cradled in cotton wool, that she had made after watching a film about a girl dying from radiation-induced leukemia. Thousands of paper origami cranes were released in her memory. I am in tears telling her about the fires of Chernobyl burning into the core of the earth.

How can I tell my daughter that this is what we are doing? I marvel at our talent for forgetting, that daily we live with this knowledge and fail to act:

> perhaps we are like stones; our own history and the history of the world embedded in us, we hold a sorrow deep within and cannot weep until that history is sung (Griffin 1994: 8).

How can we sing our history?

I remember again about the women dancing and a story Catherine Ellis, an ethnomusicologist, told at an address at the University of New England. Ellis had been recording songs from Antikirrinya women in the desert for some years during the fifties, and had built up a considerable respect and reputation with the women as a knowledgeable woman of song. When one of the older women died she had been given a particular song cycle to look after because they said there were no women alive who knew the details of this songline. Ellis decided to do some research and followed the songline to an outback town where she discovered that, in fairly recent history, a group of women had been shot by local police for performing the ceremony. As she spoke, it was clear that she was irretrievably impaired by this discovery. What should she do? What relation could she now have to the music she had spent her life collecting and recording?

~

The only writing I have among my memorabilia from Pine Gap is a text of the meeting in which Aboriginal women who supported the camp made the reasons for their support clear:

We want this land. We want to look after it well. We want to smell the good winds, not like at Maralinga when we smelt the black dust coming from there. Lots of our relatives died at Maralinga from black dust. We want all these hard thinking people to go away. We're not rich people like these people. Not strong people, we don't have power but we want to be heard. We want our children to be protected to look after our law. Old people have given us this land to look after and it's not for the war. People want to live here quietly. The Americans should fight the war in their own country. We understand that they drop one bomb and the enemy retaliates and it keeps on going on and on. The whitefellers' country is overseas, ours is the dry country in the middle of all that water. Our dreamtime stories extend to the north, south, east and west. The whole country is sacred.

These women spoke of their direct experience of radiation poisoning, of their attachment to the land, and of the songlines which map the whole country, north, south, east and west. They say whitefellers' country is overseas. Do they mean us? What responsibility do we have in the black winds, in the death and destruction of a culture? Singing our stories is a complex business and there is a great sense of responsibility which comes with this knowledge.

The next day, as I walk back from the library, a refrain echoes insistently in my head; *I do not want to live in the house of the Father, nor do I want to inhabit the Father's story. But his house is my house and his story my story.*

I worry about the problem of not being able to tell my father's story and even less the story of my father-in-law. Is this a betrayal of family loyalty? Who owns a story and who can tell it? There is a great silence about his father and his father's life. But in a very important way their stories are my stories, the stories which have shaped my life. It is only by telling these stories that my story can be changed.

this crisis of memory and forgetting is both personal and cultural. The depths plumbed in the process of recovery, in reaching for the putative lost fullness of the mind, will be as much of the psyche as of the civilisation. There is a deep link between the condition of the individual and the condition of the mass, and healing or rebalancing that remains external may not be healing at all (Brooks 1991: 36).

In *Writing a Woman's Life*, Carolyn Heilbrun talks about the time of the great dismantling, when women began to tell their stories. She refers to a generation of poets, born between two world wars (my mother's generation), who broke through the narratives that control women's lives. Denise Levertov, Jane Cooper, Carolyn Kizer, Maxine Kumin, Anne Sexton, Adrienne Rich and Sylvia Plath.

> There will be narratives of female lives only when women no longer live their lives isolated in the houses and the stories of men (Heilbrun 1988: 47).

At Pine Gap, in a women's-only space, we begin to tell new stories to each other. In telling my stories of Pine Gap now, I talk with other women—friends, colleagues. It is with, and through them, that I can speak these new stories.

> I do not believe that new stories will find their way into texts if they do not begin in oral exchanges among women in groups hearing and talking to one another. As long as women are isolated one from the other, not allowed to offer other women the most personal accounts of their lives, they will not be part of any narrative of their own (Heilbrun 1988: 46).

And yet, even for three weeks at Pine Gap, a women's-only space is highly contested. It was a complex negotiation for me to leave four children with ramifications for many years after. While tribal Aboriginal women appeared to confirm a women's space with their Two-Women dreaming dance, urban Aboriginal women had chosen a man as first speaker at the march. An ongoing debate in the women's movement was, and continues, about the right of women to women's-only spaces.

And yet, I remembered the ease with which we established a women's centre at Papunya in the old hospital building, and the hours spent in the women's camp, talking about the dancing (talking about) country. I later read Bell's account of the *tjilimi*, a culturally designated and accepted women's-only space. And out in the desert, the women singing and dancing an independent women's relationship to country that operated as a power base from which to represent themselves in the landscape, to tell their stories. I had never experienced this within my own culture, nor did I experience it on my return from Pine Gap.

16th May

I dream that I am living in a big house with my sister and her three babies. I am looking after Thomas, the eldest, and he is vomiting over and over during the night. He has vomited all over me, I feel the wetness through to my skin. In this dream I am lying on my back and my body becomes the hills and rivers, mountains and caverns of the landscape.

On waking I am immediately present to this writing. The excess of spilling over, the libidinal flow, forms the ground or fabric on which this writing grows. The ground is the 'thetic' body at its surface, a space between inside and outside, a margin. As I work on the various bits, some are more developed than others, more thought out, more abstract. These are excrescences, rocks, hills and mountains, rising above the ground. Then other bits dig deep below the ground, hollowing out caverns and tunnels, the place of dreams and poetry. Some of the underground bits, the caverns remain untold, unexplored. There are still some great gaps, fissures of possibility.

After the Two Women Dreaming dance, when we had moved out to the gates at Pine Gap, the Aboriginal women continued to participate in the camp, but they stayed elsewhere and were very noticeably absent from the site of our actions. They had their own agenda and understanding of the protest and although they met with us from time to time, a serious confusion developed. In the original meeting, with

Diana James as interpreter, Pitjantjatjara women had made it clear to us that they had powerful interests in this land, that they understood the reason for our protest in their experience of Maralinga, and that the meaning of this country is mapped already by extensive dreaming stories. Given such a strong and clear statement of parallel involvement, the events that followed were surprising and painful and remained unresolved at the end of the camp. The confusion seemed to stem from a meeting in which an Aboriginal woman had acted as interpreter, followed by rumours and counter rumours, leaving white women stranded on an island of misunderstanding. Shortly after our return I wrote to Diana James and she replied:

Dear M
The saga over the Aboriginal women's involvement in the demonstration has not been concluded. That woman Alison Hunt deliberately misinterpreted at a meeting between Aranda and Pitjantjatjara women at Telegraph Station. She then misinterpreted the important meeting at Roe Creek between Aboriginal women and Pine Gap demonstrators. She managed to create an atmosphere where Aranda women were criticising Pitjantjatjara women and the Pine Gap women were totally confused, hurt and angry.

Alison said that Nganyinytja [leader] had said she was sorry to have been involved in the demonstration because the actions of white demonstrators breaking in to the base and being arrested had brought shame on Aboriginal women who had to live there always. She continually repeated what 'idiots' the women at the demonstration were. Her translation was so obviously biased that one woman (who would not have understood Pitjantjatjara) stood up and accused her of not representing all her people's views. She [Alison] then left centre stage in a huff. I could then stand up and re-translate Nganyinytja's message which was essentially that she was glad she had come to support the women's demonstration for peace and disarmament, she would have preferred more sitting down and talking out the issues with both sides of the people involved, she didn't like the direct action taken that resulted in arrests, but she understood that this was white people's way of acting against their own wrong laws (i.e. laws that protect a place like Pine Gap).

Anyway the argument was long and hard around the traps as to whether white women had misled Aboriginal women, had given enough information or had behaved wrongly. Some Aboriginal women became very upset, Di Lane is now out west trying to sort out the mess. Hopefully it will be cleared up and this Alison woman will desist. It seems likely that she was paid or supported by the police or others interested in disrupting the demonstration. She had very sophisticated recording equipment and acted in a manner very unusual for a bush Aboriginal woman—particularly supposing she was alone.
Love D (James 1984 pers. comm.)

Lloyd, in a later analysis, suggests that:

> the diverse reactions and responses [by Aboriginal women] which challenged and questioned the very nature and principles of the women's protest was a reflection of the range of interests and political practices held by Central Australian Aborigines which do not fit neatly into one political theory and practice (Lloyd 1988: 44).

She points out that the performance of *awulye* and *inma*, women's songs and dances, can be read as Aboriginal women's response to the dynamics of the women's camp and the debates over issues of race and gender. But the size of the forum, the spontaneity of the women's ritual performances, and the reticence of the Alice Springs women to translate the ritual performance meant the finer distinctions being made by the Aboriginal women were lost on the audience (Lloyd 1988: 50).

We are dependent on the skills, goodwill and integrity of interpreters to carry us over into meanings embedded in another language. The complexities of translation. And we could be excused in giving a slight ear to the conspiracy theory that the mistranslation was instigated by Federal Police who might have had a vested interest in creating conflict between Aboriginal and non-Aboriginal women. One of the most profound and potentially powerful elements of the protest was the joint presence and involvement of hundreds of white women from all over Australia, and hundreds of Aboriginal women from all over the centre. Whatever the story, a simple explanation of mistranslation is too simplistic to account for the inability of white and black women to speak with each other. It was already such a very complex interaction, the Aboriginal women trying to understand the actions of the white women without even sharing a common language, and the white women not even able to begin to understand the complex layers of meaning about relation to land conveyed in the Two-Women dreaming dance which was performed at our first contact.

~

In analysing the experiences now, it is clear that Pine Gap opened up a space of possibilities, a liminal space which continued for many of us for many years. For me, it was the Aboriginal thread that continued into my later work: within three years of Pine Gap I had begun working on *Ingelba* (1990) with Patsy Cohen, followed by *The Sun Dancin'* (1994) and 'In Search of the Queen' (1995). It is clear now that the Aboriginal women did, in fact, present the first tissue paper layer of a different level of mapping the land with songs and songlines; that the disruption of straight lines and roads by circles of hats and parasols opened up possibilities for new meanings to emerge and that those possibilities are still open. And my embodied three-dimensional mapping of hills and rivers, valleys and caves, is a sand map written on the body that, like the dreaming, can always be recast.

Performance II

Two Women Dreaming

Landmarks
Angatja, *July, 1993*

Each morning we open our eyes to an ever changing sky through patterns of feathery mulga leaves from an envelope of green canvas, a fire of mulga at our feet, and curved mulga branches at our head. The camp is a clump of mulga trees beside a little hill, a huge pile of reddish brown rocks, *puli*. From the top of the hill miles of desert landscape is laid out in every direction, only the odd sleeping rock breaking the even pattern of green dots on red. Far in the distance I imagine I see Uluru on the horizon to the north. As far as we can walk in any direction it is the same, red earth scattered with desert flowers and shrubs and occasional clumps of mulgas. There are twenty of us dotted through the mulgas and the mornings are cold. As we wake, we make our way to the kitchen, a big open fire with a curved wall of intertwining branches. Each morning Nganyinytja comes from her camp, a little way away, to talk to us about her life, and how we will spend our day. This is the pattern for the eight days we spend here. We are in the middle of Pitjantjatjara Lands, six hundred kilometres south west of Alice Springs.

In this performance I return to the desert with Nganyinytja, a Pitjantjatjara woman who runs an ecotourist enterprise to introduce white people to her country. Nganyinytja believes that *anangu*, Pitjantjatjara people, 'have always kept our land and looked after it and make it grow' (Macken 1993: 23), and

*Nganyinytja at ka_ltu ka_ltu, 1993
(Margaret Somerville)*

if she can pass this knowledge on to white people the land and its people will be healed. In this I have access to a world of meaning that was not available to me in my earlier dancing in the desert. Without someone to translate, to provide a bridge for me to access this other world, the experience is mute. Here song, music, songline, dance, people and place come together under Nganyinytja's careful and caring teaching. In this performance I ask how can I move across this space between Nganyinytja and myself. I use Elspeth Probyn's analogy of relations between self and other as a fold through which the outside is folded in to the inside and ask how can I do this without self-annihilation. Every cell of my being wants to absorb this place and it is imagination that carries me across the abyss—'imagination is not a luxury but a lifeline' (hooks 1991: 55)—and language that holds the meaning of place.

~

Elspeth Probyn writes '[e]ntering the imagination of others requires':

> the reader shift her paradigms ... This may indeed require them to relinquish privilege and their acceptance of dominant ways of knowing as preparation for hearing different voices. The ability to be empathetic is rooted in our capacity to imagine (hooks cited by Probyn 1993b: 148).

Tjukurpa ngaranyi puḻingka munu iwara, ankutja, tjukurpa ankutja.
This is a story of rocks and tracks.

We leave Alice Springs squashed into a tiny bone-rattling bus, and seem to be forever going, going, going, *ankula ankula ankula*, until I feel like I'm in a dot painting myself. Dark red dots of a newly sprouted desert succulent, clumps of bleached gold spinifex and, further away, dots of grey mulgas on a background of red ochre pass before my eyes.

Finally just when I can't sit on a bus rattling through corrugations of dirt any longer, we stop at a small hill—red blood, the colour of ironstone—and Nganyinytja leads us a little way up the hill to a flat rock surface with a rockhole. A deep triangular slit in the red rock, filled with water, reflects the brilliant blue of the sky and a luxuriant growth of feathery green grass protects the slit at one end. The surprise of water, and of blue in this arid landscape.

Nganyinytja then draws our attention away from the rockhole to the distance, a vast stretch of desert landscape ringed by purple blue hills. She circles this vast expanse and gestures with outstretched arm and chin, as she says *this is my front doorstep.*

We arrive at our campsite at Angatja and are swarmed by wild straw-haired kids chattering in harsh gutteral voices. *Ngalya kati*! bring it here, they shout as swags are unloaded off the truck. I call back to them in their language and they fall about laughing at me. One bright skinny kid comes to help unpack, undoing toilet bag, putting on make up, examining clothes and tape recorder. We exchange names; *Margaret*ta she says, adding her own rhythms. No adult appears until Nganyinytja wanders into our camp, bare feet, flowered skirt, and cardigan wrapped close against a chill wind.

Palya, Ngalya-pitjala nyawa ngayaku ngura
Welcome, come and visit my country, she greets us.

Each day after that, she comes in the morning and again in the afternoon.

Nganyinytja cradles the whole experience in her life story. This is what she gives to us and what she holds us in. She remembers as a small child wandering through the vast stretches of country to the west of Angatja towards the border of Western Australia with her mother and father, before white man came to her country. She tells us how they saw the first white men come to Angatja on camels from the top of the same hill where we are camped. Their mothers had hidden all the children in the rocks on the hill, terrified of what might happen to them when the white men on camels arrived.

Who is this small grey-haired woman who comes to us with white floured hands from making damper? I see her as a vision of the five black matriarchs from my work with Patsy Cohen on *Ingelba*; the embodiment of Mary Jane Cain of *The Sun Dancin'*, the woman who straddles two eras of history—the time before white settlement of this land and the time after. She moves between two worlds of such profound difference, and she gives her people the strength to move forward. *We gotta make it good for ourselves to go forward*, the people say.

How can I move across this space between Nganyinytja and me?

~

Monday, 23rd August, 1993

It is just over a month since I returned from the desert and there has been nothing but illness. No words, only pain. I could politely explain it as the flu. Perhaps it is. In the desert Gretta and Barbara were both sick with the flu and I seemed to come down with it in the last two days in Alice Springs—sore all over the surface of my skin, aching joints, feeling sick (at heart) ... Before I left for Alice Springs, Lyn had said that Aboriginal spirituality is too powerful, she chooses to leave it alone.

In Brisbane, on my way home, I felt madness in the pain of moving between too many worlds. It took seven days and seven nights to get home—Angatja to Uluru, Uluru to Alice Springs, Alice Springs to Brisbane, Brisbane to Armidale—seven days and seven nights since I had my photograph taken with Nganyinytja, smiling in the desert.

I have three flint stones lying in the front pocket of my backpack. I rehearse their stories. The first two I picked up at

Pantu, a salt lake created by the sweeping of the *ngintaka*'s tail. 'We came here as children, in the olden days, *iriti*', Nganyinytja says, 'and camped here so our parents could tell us the story.' She bent down and picked up the shining white flint blades from the side of the lake to show us that they had indeed been there. The third, Nganyinytja casually picked up from the ground as she told the women about cutting the umbilical cord during childbirth. 'We cut the cord with this, *puli*,' she said, and handed the stone to me.

At Uluru it was said that people had come back with stones because they had got sick after taking them away. As I become more ill I imagine placing the flint stones in a padded postpak and sending them to Nganyinytja in the desert. But it is the same disjuncture as with the women I had danced with earlier; the camp at Angatja doesn't really have an address, and Nganyinytja can't read, so how can I explain to her that I have to send the rocks back because I am frightened they are making me sick? I am too ashamed at being just another stupid white tourist who doesn't know how to behave.

Also lying in my bag are two tapes of Nganyinytja's stories. I leave them there, unable to even imagine beginning the task of transcribing Pitjantjatjara. Transcribing is difficult at the best of times but how to cross the gap of another language?

On Saturday night at Diana James's[1] place in Alice Springs before we set out for Angatja, I am curious that we sit around an open fire in a suburban backyard while she tells us the history of Desert Tracks. Is this just an open fire transported from the desert, or is it an easy way to entertain big mobs of people, or does it represent something of an in-between space for her, a space in which to negotiate the crossing over between two cultures?

She tells us that Nganyinytja has been interested in taking white people to her country since the 1940s. In the 1980s a

---

1 Diana James co-owned the ecotourist business Desert Tracks with Nganyinytja at the time of our visit. I am indebted to both her and Nganyinytja for this story.

group of people involved in the Pitjantjatjara landrights case came through on an organised tour. A little while later in 1988, Diana, as interpreter and business organiser, and Nganyinytja, as teacher/leader, began Desert Tracks. Barbara, the camp cook, now on her third trip to Angatja, talks about how much some people change after being with Nganyinytja in the desert. She tells us about a high flying business executive who wrote a poem about his experience, and she says 'the closer the experience is to the earth, the greater the transformation.'

> Recalling how for Foucault, the processes of subjectification are the ways in which the exterior line of force is bent and folded upon itself so that the inside and the outside are rearticulated, I can then argue for the self as designating the ways in which the experiential is bent upon itself (Probyn 1993: 5).

In the desert I live on the ground in a canvas envelope called *kaltu-kaltu*, after the main grass seed that the women used to make flour. My head is protected by a *wiltja*, a small curved structure of interwoven mulga branches; at my feet a mulga wood fire smoulders slowly to keep away night spirits, and above I look through feathery leaves of mulga at patterns and colour of sky. One morning I wake to wisps of pink against the palest of pale blue skies; on my left the *puli*, the rock hill, is lit with an orange luminesence by the same caress of pink. I close my eyes and minutes later awaken to uninterrupted, vivid blue of day, the hill now wearing its day shade of desert brown. Things are transformed in the desert.

In the desert there is an ants' nest in red dirt surrounded by an arc of tiny purple petals, a fragile china blue against desert red. The ants are busy taking one tiny petal at a time into their nest to gather the small droplet of nectar at the base of each petal. Once milked of nectar the ants bring the petal out again and deposit it in the growing circle of petals that decorates their doorstep. The colours belong to the daisy flowers of a small desert plant that blooms briefly after rain. After rain the red dirt comes to life. Seeds sprout, green leaves appear and flowers bloom until the desert is a carpet of green splashed with brilliant colour. Caterpillars proliferate and ants are busy everywhere. The next day we return to photograph the brilliant purple blue fluorescence but the petals have completely vanished. The cycle is complete.

In the desert, John the lawyer keeps on asking where did the law begin? Ron the translator is having problems with his questions. Nganyinytja becomes agitated. The *law* she says, *tjukurpa*, the dreaming, has no beginning and it has no end. He repeats the question in a slightly different way. 'But where does the law come from?' he says. She touches the ground where she is sitting, '*Manta*', she says, 'the ground, it comes from the ground. You can see it on the ground, in the hills, the stories, the stories are everywhere, they have no beginning, they come from nowhere, *they are in the ground.*'

Tuesday, 24th August

Back in Armidale the naturopath says your body is in chaos, you have no centre. He prescribes a detoxification diet—clean water, clean raw foods, no tea, coffee, alcohol or cigarettes. For five days and nights I don't care whether I eat or not and days merge into nights of waking sleepfulness or sleeping wakefulness. All night I dream waking, confused loops of thought dreams—about the desert, Alice Springs, Nganyinytja—as if I am connected to that place by countless threads of thought patterns that seem to have become tangled into an impossible mess.

> One way of imagining the self is to think of it as a combination of acetate transparencies: layers and layers of lines and directions that are figured together and in depth, only then to be rearranged again (Probyn 1993b: 1).

In this I return to the layers of tissue paper mapping of Pine Gap, but this time I begin to insert myself into the layers as Probyn (1993b: 1) describes the process, 'a discursive arrangement that holds together in tension the different lines of race and sexuality that form and reform our senses of self.'

In the desert there is no writing, only sound. 'Sounds are bubbles on the surface of silence' (Trinh Minh-Ha, 1992: 4). I/i am stripped of everything that is me. I have no home to close the door on the world, no friends for cappuccinos, no car, no familiar landscape and even my few clothes are in another pack for space. Structureless. Someone asks me my name. Maggie I reply, changing it. My ears strain to hear another language. Every cell of my being wants to absorb this place.

The black ceiling in the Bistro closes in on me. Inside my office, my house, I feel like a mole, blinded by the light, wanting to crawl back into my winter burrow and hide. Perhaps I am in a cocoon, a chrysalis waiting to emerge. I think of the lightness of the desert and begin by eating my breakfast in the sun. I feel the sun shining on the middle of my forehead, the pineal gland, wakening me to the warmth of spring.

In the desert there is a great expansiveness, a lightness of being. I breathe deeper, climb the highest hill and see as far as the eye can see. From the top of the hill I can see for miles in every direction, as far as the Rock on a clear day. From here I learn where and when the sun rises and sets in this place, I orientate myself in relation to all the places we see on the ground. Am I learning to see? Learning to get my direction in this country?

In the desert, sleeping in the open in my canvas envelope is a challenge, but usually a calm one of being in touch with stars, moon and whispering of wind in trees. But there is one particular night when puffs of wind stir up the spirits and agitate the mind. The wind begins flickering playfully, eddying in small dusty gusts which become stronger and wilder as the night goes on. Each gust wakes me with a strange and eerie feeling, and the sleep in between is increasingly shallow and agitated. Finally, in one huge crescendo at about one in the morning the wind roars through the camp, carrying on it the plaintive howling of a pack of dingoes in nearby hills. Light spots of rain. I have never experienced this wind anywhere but in the desert. Someone has died, I say to the wakeful, and fall into a troubled sleep until morning, worrying for Nganyinytja. A grey and sombre dawn greets us and Nganyinytja doesn't appear as usual at our breakfast. Tarps are raised and swags placed under them. As we return from breakfast a soft wailing rises up from the *anangu* camp. A messenger comes to tell us that a police aide has been killed in a car accident on the road to Uluru in the middle of the night. On Friday the people will go to Pipalyatjara for the funeral. Morning plans are cancelled. Living and dying in the desert. In the desert it is said that the wind blows to whip away the tracks and release the spirit of the dead.

I learn to listen to the spirits of this space in between.

I remember one very clear dream from the desert of Nganyinytja and myself sitting on the top of Uluru, in the centre of the world. Half awake, I puzzle about how we got there because the only track up the rock is the sacred track of the haretail wallaby dreaming. All the tourists who struggle up and down the Rock become little wallabies. The image of myself and Nganyinytja is transposed with the photograph of me sitting on top of the hill beside our camp.

> I-have-no-centre.
> How does one get a centre?
> A journey to the centre.
> Which centre?
> The centre of the country
> or the centre of me?
> and what will I find when I get there?
> what if I find nothing
> a great black abyss?
> How to make the threads
> across the nothingness
> of the space between us?

Relax, deepen and *centre* yourself they say at yoga. The physiotherapist and Feldenkrais practitioner describes movement from a *centre* behind the solar plexus, the belly button, moving from a core. The yoga book says this is *manipura chakra* and that *manipura* means city of jewels, fire centre, focal point of heat and lustrous like a jewel, radiant with vitality and energy.

> The *manipura chakra* is the centre of vitality in psychic and physical bodies, where *prana* (upward moving vitality) and the *apana* (downward moving vitality) meet and join each other, generating the heat that is necessary for supporting life (Saraswati 1973: 333).

I practise relaxing, deepening and centering myself.

> up to what point can we unfold the line without falling into an unbreathable void, into death, and how to fold it without nonetheless losing contact with it, in order to constitute it as an inside co-present with an outside, applicable to the outside? (Deleuze 1990: 153 cited in Probyn 1993b: 129).

Wednesday, 25th August

After five days and nights I no longer think I will die. I begin to feel hunger and a tiny energy source like a flame is kindled deep in the centre of my being, behind the solar plexus, the manipura chakra. I am still very weak, but I feel clean and clear, a medium or conduit ready to process a message.

> In simple terms this is to acknowledge that we need to push our selves and feel our selves moving ... The point is not to include the outside with the inside in such a way that overwhelming pain renders us only able to inhabit the inside. Folding the line reconstitutes us in another form of subjectification; it does not annul us (Probyn 1993b: 129).

I still cannot be in my office, there is too much pain and I am afraid of isolation.

At home I see the clapsticks on the table and want to sing and dance with all the people as we did nightly in the desert. Trinh says:

> *Every illness is a musical problem*
> Music has a magical, energizing and creative power. The mere shaking of a cowbell is enough to make people drift into a state

of excitement. It is then said that 'strength has entered them'. Elders who can hardly move in daily situations without a cane would emit war cries and dance frantically to the sound of music. Farmers who feel tired and lack enthusiasm will be fired with desire to work in the fields upon hearing the drum beats or the chants of the masks (Trinh 1992: 7).

In the desert, each night after we eat in our camp, we hear the sound of singing calling us on the night air. We walk hesitantly in the dark, guided by feel of earth under foot, to the light of fires dotted through the dancing place. Women, men and kids are already sitting on the ground in the firelight, in their separate groups. *Ngalya-nyina*, sit down, the women say, banging the ground beside them. The singing rises and falls spontaneously to the thud thud of the beat, hands on thighs, cupped hand against cupped hand or stick on ground, rhythm and counter rhythm that runs through the earth and up into our bodies.

> One of the most important points that must be understood about traditional cultural knowledge in Central Australia is that it is centred on song knowledge. The definition of a knowledgeable person is the person 'knowing many songs', for without song knowledge, information about places, laws, correct behaviour, healing, food sources and a host of other items is unavailable ... rights to land and relationship to country are articulated in the performance of song series or songlines relating to the travels of the Ancestral Beings through the country in the Dreaming times (Ellis and Barwick 1989: 21).

A babble of voices rises in between songs as texts and song-lines, singers and dancers are negotiated—the honey ant song, the *ngintaka* song, the seven sisters dreaming. After one night I can beat with the main rhythm; after two I can add the counter rhythm as well and by the end of the week I can sing with some of the songs even though I can only imagine their meaning.

The ability to imagine is not necessarily couched in pain: re-membering personal experience is, however, a precondition for the capacity to articulate rhizomatic lines that touch and connect with the aspirations of others (Probyn 1993b: 148).

In the desert we learn to dance by firelight, following the body movements of *anangu* dancers. The men learn first, the emu dance, temporarily transformed, much to the delight of the *anangu* men, into head-jerking foot-stalking emus. Then the women dance in a row across the dark ground. We are the women in the *tjala* honey ant dreaming, swinging our digging sticks from side to side as our feet move two steps this way and two steps that, carefully making the story marks on the ground. Then we scrape the ground in little flat jumps towards the singing faces and collapse in clouds of dust and laughter as we finish too close to our audience.

> To imagine then, was a way to begin the process of transforming reality. All that we cannot imagine will never come into being (hooks 1991: 55).

In the desert we learn about the negotiation of gender arrangements through dance. After both young and old *anangu* women dance with bare breasts, there is much discussion in the white camp. Dancing for women is more powerful, more correct, if the breasts are exposed and painted with the ceremonial designs. Narelle, radical feminist, is loudly opposed to exposing her breasts in front of any white male. Judith, a mother, says that she thinks it is really important to dance bare-breasted. Gretta, her daughter, will not bare her breasts, outraged that her mother would even consider it. The white men claim that the sexuality of breasts is not an issue in this context but Narelle argues that it is their conditioning to view breasts sexually. Ron, the pastor, is opposed on religious grounds. And so the discussion goes on.

Eileen introduces herself in bawdily suggestive terms to Ron, pastor/translator. A large woman, perhaps in her fifties, blind in one eye from trachoma, she stands unusually close to him when he asks her who she is, and she says in Pitjantjatjara, 'I am a widow, and old but I am still strong and flexible, good for dancing and for digging ...' At the dancing, she dances the honey ant dance, large breasts swinging with the shining honey ant design, new acrylic silver against brown skin. When she returns to her seat on the ground, instead of facing the dancers, she faces the audience, and especially Ron the pastor, breasts forward. Narelle calls out 'Can I take a photo of you Eileen?' Whereupon Eileen laughs and lifts her large breasts further, an object for the photograph.

On the final night at the dancing bare breasts is on the agenda. We have become bolder and there is much teasing and suggestiveness between *anangu* and white women. It is eventually agreed, through a combination of Pitjantjatjara and mime and much laughing, that the white women will dance bare-breasted. But Ron, the pastor, intervenes. He speaks to the *anangu* women in Pitjantjatjara, but they override him because they now have the collusion of the white women. He begs the support of the *anangu* men. 'It is cold and late', he says in Pitjantjatjara, 'and it will take a long time. It is dangerous because one of the young women has been sick.' It is clear that the men are not too enthusiastic in their support but he is *tjilpi* (old man), therefore they respect his wishes, and they have known him a long time. The debate continues for some time and finally it is resolved, more by a lack of resolution than a conscious decision, that we will not dance bare-breasted tonight. 'Kuwari (later),' Nganyinytja says. But this is our last night.

I learn later that Ilyatjari, Nganyinytja's husband, has gone to great trouble to ring Diana in Alice Springs on the following day. The *anangu* men had deduced, correctly, that Ron was embarrassed by the exposure of breasts. They considered this not an adequate excuse compared to the importance of proper dancing. They rang to tell Diana that he should not be used as interpreter anymore. It is critically important that the white women dance bare-breasted, they say.

Diana tells us it is important to learn to sing and dance because that is how to learn the culture. Ilyatjari says that it is important for the white women to dance *so that they become part of the re-creation of all those mayi* (food plants). The performance for him not only functions to teach the white people about their country, it serves the traditional purpose of increasing the plants which produce the grass seed in the *ngin̲taka* dreaming.

> Imagination can enable us to understand fictive realities that in no way resemble where we are coming from ... to enter realms of the unknown with no will to colonise or possess (hooks 1991: 57–8).

Thursday, 26th August

I take Nganyinytja's tapes out of the front pocket of my red backpack. The language is definitely one of the barriers of pain. At first I transcribe only the English words and hear very little Pitjantjatjara. When I was in the desert before, there was no interpreter, I could only understand the language I knew myself. This time, because there is some translation from Pitjantjatjara, there are now words and stories.

In the desert Nganyinytja always speaks to us in Pitjantjatjara, but there is a male translator. When a question is asked in the mixed group about childbirth, she says that she cannot speak about it in front of men, it is women's business, and she will take the women off to the bush to talk about it by herself. The women go with her on the last afternoon, and, with a mixture of mime, her very little English and my very little Pitjantjatjara, we communicate more closely with Nganyinytja as a group of women than we did at any other time.

I wonder about Nganyinytja's refusal of English and think that it is a remarkably strong gesture of resistance. She has clearly had many opportunities to learn—as a young girl in the newly established school at Ernabella she was the first pupil to become a teacher, according to Ron, 'teaching the younger children what she had only learned herself a few months ago'. She has participated in countless discussions and meetings over the years with Government officials, lawyers, other Aboriginal people over land, land rights, and so on, and always carries out these negotiations in Pitjantjatjara with an interpreter/translator.

I realise that translation/interpretation is an inevitable and fundamental component of Nganyinytja's vision and brood about the question of translation: translation is a movement across, 'with etymological and semantic connections with metaphor, transfer, transference and transport' (McDonald 1985: 94). According to Eugenio Donato, the German word for translation, *ubersetzung*, has a double etymology and thus a somewhat stronger semantic field, since the other sense of *ubersetzung*, meaning 'translation, metaphor, transfer', is 'to leap over an abyss' (McDonald 1985: 127).

In the desert, each day when Nganyinytja comes to tell her stories and Ron arrives to translate them, she and Ron renew their connection, and their collaboration. They sit side by side in the circle of our campstools, talking comfortably in Pitjantjatjara, 'How did you sleep?,' they ask each other, 'How is Ilyatjari?' (Nganyinytja's husband has been sick); they talk about the weather, and what the day will bring. They have known each other for fifty-three years since Nganyinytja was a girl and Ron was the first teacher at the Ernabella Mission School, and they have great pleasure in making stories of their mutual memories. Nganyinytja had asked for Ron to be the translator on this occasion.[2] Sometimes Ron translates (almost) word for word:

Tjukurpa ngaranyi puḻingka munu iwara, ankuntja, tjukurpa ankuntja.
It's a story of rocks and tracks.

Sometimes he gives extra information about the words he is translating:

---

2 The translations are as given by Ron and recorded on tape.

The way she describes it is 'we used to have our babies on the ground'—those are her words, *panangka* means on the ground, *pana* means ground.

Sometimes he gives a general interpretation of a section of the story, adding his own memories—and his values:

> Now this is significant, I didn't know she was going to come on to this, she had imbibed so much trust of the medical system that we had. She happened to have a very good relationship with Dr Duguid and his family, and so when she went to Amata, to this new community ... she was able to persuade the people to entrust themselves to white man's medicine. She was the first one to have a baby in the hospital. See, none of them would ever go near a hospital to have a baby. She used a word that is hard to translate, it means imparting a trust, trustifying them if you like. She was the first one—I'm gonna have my first baby in the clinic, in the hospital—which none of them had ever thought of, they always used to have it outside.

And often he adds stories of his own:

> By the way, one of her contemporaries—the school had anything from two hundred in one day to ten another day, it was so much variety, they were all nomads you see—but she stayed quite a while—but one of her contemporaries was called Tjinima—she lives in Ernabella right now—and she attended every day the first year, and all the rest of the kids were totally nomadic and came and went, including Nganyinytja.

Barbara (the cook) has been on three trips with Desert Tracks and she says that each time she has heard a different story from Nganyinytja because there were three different interpreters. It is not only that the translation is different but that the production of the story itself is a collaboration. With Diana, she says, the emphasis is on Aboriginal politics; with Linda, the focus is on the day-to-day life of the people, and with Ron, there is more of the history of the people.

We are totally dependent on this relationship between white translator and *anangu* storyteller to enable us to understand Nganyinytja's stories, and we become more and more frustrated with Ron's particular bias in translation. One day he says 'They have no abstract thought,' and I wonder, having grappled with the complexity of the layers of symbolic meaning in the Two Women Dreaming stories, what he can possibly mean. Later it becomes apparent that in his work of translating the Christian message into Pitjantjatjara, there is great difficulty translating Christian concepts of love and forgiveness into Pitjantjatjara because there are no equivalent concepts. So we deduce that what he means by having no abstract thought is that the *anangu* didn't have the same abstract thought as him.

We try to negotiate with Ron, a translation closer to our own values and understandings: Narelle asks, on the question of childbirth in the hospital, whether some of the changes might have been destructive. Ron answers yes, some were, and gives the example of drunken violence. We miss the point with each other; some of us feel that the movement to childbirth in hospital, or even the introduction of Christianity, might be destructive as well.

Not only are we dependent on Ron's translation but we cannot speak with Nganyinytja directly ourselves. John, the lawyer, asks Ron to ask Nganyinytja how her people can show such extraordinary generosity of spirit after what has happened to them. Ron answers on her behalf that 'The present generation doesn't know anything about that, they don't look back on that, she didn't suffer from white man's deprivation.' But John insists 'Much of it is still part of of their cultural experience which those people had, could you ask her that question, how she finds that generosity?' Nganyinytja replies that it is partly due to her Christian faith, partly due to the fact that in her family there was a feeling of generosity and affection for other people, and also because now they have their land back she is at peace with the world and wants other people to share it because it is such a lovely part of the world. John then asks do they have a word for bitterness or resentment and Ron replies only *pikaringanyi*, to become angry, and asks Nganyinytja, does she feel anger.

There follows a long and interesting discussion between Ron and Nganyinytja which I shall summarise, using Ron's translation from the tapes, as follows:

> She said that her family previously were surrounded by a lot of fighting amongst the people themselves (in the drought times), killing one another, and all that, and the whites did the same thing. She mentioned Kenmore Park, and when I came, Kenmore Park—the owner there, he would kill any black person he saw, he hated them. And she said she was brought up with the environment of anger itself. Then the mission started, and it wasn't just a mission, it was a buffer state. It was a place where they could find refuge, an in-between—in-between their drought stricken life out there and the white life in Alice and places, which wouldn't have been good for them. It was a haven and she said, 'We began to see that we could live a life of peace.'
>
> She then felt, which I imagine would be quite unique, a feeling of good will towards the white people, that she hadn't experienced before. So when we started as a—not only a mission but there were government people here—to keep people out, which we're still doing, well she thought I don't feel like that about these people, it would be lovely if we could show them good will.

So according to Nganyinytja, the people were caught between the intense tribal fighting of a severe drought and the killing of the early white settlers, and the Mission provided an in-between space, a haven in which they could begin to accommodate change and tell their new cultural stories. Later, when the land was defined as Aboriginal-owned and outsiders were not allowed access to Pitjantjatjara lands, Nganyinytja felt that she wanted to invite outsiders in so they could learn about her country. Ron described her as 'unique' in her attitude to white people.

Without Ron, the translator, I would know none of this.

At night, in the dark, we walk together with Ron to the dancing, he lighting the way with the torch and us carrying his folding chair, joking as we go. We have come to love the 'intelligent patriarch'[3] who loves difference, grew up in Egypt, raised eight children and buried his first wife in the desert. He left the desert and retired to England but was called back by tapes from the *anangu* to live out his life in the desert. At the dancing Ron sits towards the back, distant, with a blanket over his legs. Sitting on the ground we are warmed by the little fires dotted all around. At first, in the space in between the small songs, Nganyinytja tries several times to call out to Ron, over the babble of talk and laughter, the text of the song, but Ron does not translate. He says translating the song text is too difficult. I wonder, is this because of the noise, the untranslatability of the texts of the songs, or his difficulty with this aspect of *anangu* life? Nganyinytja stops trying to tell the meaning and relaxes into the singing.

Who has power, who chooses what is spoken into existence? Do I want to be, can I be, carried across the abyss by the 'intelligent patriarch'?

In the desert, Dennis asks Nganyinytja why she originally wanted to have tourists come to Angatja. Nganyinytja replies that when she saw all these people being sent off[4] and they had to continue their lives in ignorance of all the beauties and wonders of life here, she compared it to the times when the white people would send off and resist the blacks. She had a feeling of love and sympathy for people who wanted to see this but couldn't, so she felt they should provide an area where they could invite them. Dennis then asks whether it was good to get the income and Nganyinytja immediately replies:

pitjangku kulintja wiyala, mukulanguru kulintja, alatji,
it was love, not money.

---

3 After Mary Bastable (1993: 74), whose story, 'The Intelligent Patriarch is Always Melancholy', introduces me to the idea of 'manipulating the patriarch' through writing, and thus 'the intelligent patriarch' is a shorthand way of naming patriarchy and the relationship between male person (penis) and language (phallus).

4 Angatja is situated within Pitjantjatjara Lands, which have restricted entry to outsiders, and virtually the only way to gain entry is through Desert Tracks.

Nganyinytja and Diana have together materialised Nganyinytja's visionary dream by operating Desert Tracks as a commercial business since the project began in 1988. Diana as interpretor/translator, cook and bus driver, has supported the venture with her labour and finances. It is quite literally a Two Women Dreaming. Why do they dream it as a business venture, now staggering under the crippling pressure to make more money and to speak and conceive of this vision in tourist dollars? Nganyinytja says quite clearly that she has no interest in the money. Perhaps it is because over and over again, *anangu* dreams—as with other Aboriginal desires—are re-spoken, translated in terms of the current discourses of the day, the discourses of power. Why is it that this—perhaps the most important educational experience in this country—is not funded as an educational activity? How much does it change how Nganyinytja is able to speak the stories of her country, that she has to frame them in terms of economic rationalist discourse? What if her vision was funded for instance, as an open university?

There are not just simply many different stories, but those different stories exist inside relations of power. In refusing English, Nganyinytja maintains the integrity of her stories, but who can speak, whose stories are heard, and by whom? Perhaps the new discourse of reconciliation enables Nganyinytja to respeak her vision in terms closer to her expressed desires than is possible in the economic rationalist discourse of enterprise:

> Reconciliation means bringing the two cultures together: *maru piranpa tjunguringanyi*, black and white coming together. The two laws need to become one to keep the land. We, the Pitjantjatjara people have always kept our land and looked after it and make it grow ... If people will listen to our way they will understand why we live in the country of our grandparents and why we must have strong land rights. If people lose their land their law is broken and their spirit dies (Macken 1993: 23).

Nganyinytja is moving across the space in-between with a vision about land, about the practices of space, in which she believes that white people can be open to alternative (post-colonial) ways of relating to land. There is a meeting of two rather than an appropriation by one. She goes on to say that:

> much trouble has come from people forgetting the land, the spirit. Many people are sick and have lost their spirit. The white government has cut their culture; we grieve for them. But we can all learn to make our spirit strong (Macken 1993: 23).

I think about my own movement across, and return to the fundamental question of language, buy a Pitjantjatjara Dictionary and begin to learn to hear the words, patiently listening to the sound of Pitjantjatjara on the tapes and growing in understanding day by day.

> I want to run the risk of speaking within the space between my self and another's self. This then entails working within a 'limit attitude', speaking with attitude, as I attempt to elaborate ways in which we can transform limitation ('I' am not 'she') into practical critique (Probyn 1993b: 145).

The threads that I need to cross the abyss are the threads of language.

Friday, 27th August

Yuwo -o! (expression of affirmation, delight after *uwa* meaning yes).
For the first time I can hear the words of a song and know what they mean—well the literal meaning anyway. I have had a breakthrough with a high quality tape recorder and much patience, and I feel as if I have been liberated from a semiotic muteness that has pervaded my experiences of women's ceremonies in the desert into a world of lightness and understanding. I follow the language of the *ngintaka* story:

> *Tjukurpa ngaranyi pulinka munu iwara, ankuntja, tjukurpa ankuntja.*
> This is a story of rocks and tracks.

Tjukurpa means word; it also means story; it also means the Dreaming, the Law; and it means the stories that Nganyinytja told us in the desert by day in the circle of our camp and by night in song and dance.

Pu_li(ngka) means rock; it is also the pile of rocks that made the hill beside our camp; and it is the ranges of small mountains that curve and bend around this country made by *wati ngin_taka;*[5] and it is the material of the flintstone knife I carry in my backpack.

Iwara is the mark made by animals, including humans on the ground as they move about their lives; it is also the track that connects one place to another, including roads; and it is the visible material representation of the Dreaming ancestor moving through the country along a songline.

Ankuntja is going journeying through country, and it is Nganyinytja's body and facial gesture travelling metaphorically through space in story as hand and fingers, face and body, map out the direction, scanning the places as she goes.

Language carries the signs of the relationship between people and place.

Uwankarangku an_angu uwankarangku, ngin_taka tjukurpa. Tjukurpa uwankara ngaranyi an_anguku mantangka uwankarangka tjukurpa kulintja ... pu_lingka tjukurpa pu_lintjaku iwaraku.
This is a story which has not been put into writing.

Ka ngan_ana inma ngaranyi, an_anguku wati tju_taku, inma ngaranyi tjukurpa nganampa kanyini kana inma inkanyi ... ngin_takaku inma
and it's a story that we sing, a song cycle.

Minymaku tjitjiku ka kutjupa mantal
and there are connected with it separate song cycles which are sacred which they don't mix like this. But what you've seen is the public one.

---

5 *Ngin_taka*—perentie lizards—are big lizards growing up to two metres in length, with long powerful necks, a wedge-shaped head, and jaws that can crunch down a large *tinka* [goanna] or rabbit. They stand on sturdy legs with long-toed feet and long tough claws. The *ngin_taka*'s skin is made beautiful by bands of spots against a mottled grey-white background. The spots down the back are white ringed with black. On the legs they can appear as white spots on a black background ... A confident woman will approach one that has resorted to freezing in grass cover. She will circle the spot, grunting a sort of chant, slowly closing in. When the creature is mesmerised, she will break its back with her crowbar or a rock (Bryce 1992: 32–3).

In the desert we hear layers of *ngintaka tjukurpa*. In our morning story sessions we sit around the circle of our camp kitchen while Nganyinytja tells us the whole story of the songline of the *ngintaka* man as he travels through the country; then we actually travel along the *ngintaka* trail and at each place singing and dancing links each part of the story to a particular landform—the great salt lake, *pantu*, where the *ngintaka* swayed his tail from side to side, the cave which is the empty belly of the *ngintaka* after he vomits. At night, in the light of lots of small fires at the dancing place, all the small dances and songs of the *ngintaka* ceremony are again performed and we learn through rhythm, song and dance.

Nganyinytja makes no distinction between telling the story of her own life, wandering through the country as a girl, and the wanderings of *wati ngintaka* through the same landscape, making the shapes of the land that she loves. She inherited rights in the dreaming through her father and, although these rights are not exclusive, she has considerable authority over the country and the people who live there on the basis of her rights to the *wati ngintaka* dreaming. Each morning after breakfast, Nganyinytja appears in storytelling mode, and waits until everyone is ready to tell the next episode of the *ngintaka* songline.

Palya (is everything OK)?
Kulila (listen)!

The ngintaka *man lived at Ataramula, near the border of Western Australia and South Australia, with his wife and they had two inferior grinding stones. Because of this they weren't able to grind their grass seed and so they just ate the seed itself, unground. One day he heard a huge grinding noise in the east so he got his spears and spear-thrower and started to walk, travelling day after day* (ankula ankula ankula). *He finally arrived at Walatina (Yami Lester's country)*[6] *and saw a lot of people there happily grinding*

---

6 See Wafer (1990).

their seeds and eating their food. He was shy and frightened but then he saw his daughter who had come there before him and they brought him some food. When he ate the bread that has been ground he thought it was marvellous because he hadn't had anything like that before, only raw seeds. He watched them from a distance and then he saw this beautiful tjiwa. *He said I've come a long way to here and I'm just looking at something that looks very wonderful to me. So they stayed a day and then they went hunting with the other men and the women went for* wangunu *(a grass seed) and they came back and slept the night. They had an inma together because of him and they had a wonderful ceremony. He pretended that he had hurt his foot and when the men went hunting and the women went for grass seed he looked for the* tjiwa *and he took it and hid it inside his tail. So then, with this big* tjiwa *inside his long tail, he started to go back west. When the people came back from hunting and foraging, that beautiful* tjiwa *wasn't there any longer. And being a* ngintaka *out of him flowed scores of hundreds of little goannas and their footsteps were everywhere. They were grief-stricken because this beautiful* tjiwa *was gone and they started to look all over and they realised that the* ngintaka *man had stolen it. A boy managed to pick up his tracks and so a whole lot of them went west and they saw him in the distance going that way. He was standing so they came and touched him all over but they couldn't feel anything. They kept on following him and feeling him to feel what had happened to this* tjiwa *and he was singing about it so they sang the story of that adventure and it is the* ngintaka *song cycle that you were hearing last night.*

They gave up after a while and all went back so he came back this way the second time. He kept this tjiwa *inside his tail the whole time and he came here to Angatja and made a* wiltja *(bush shelter) over in that direction. And they followed him there and touched him all over and couldn't locate it. (Wherever he went that explains that geographical phenomenon.) And then he went to Tjanmata and you can see where he perfected and placed the rocks at Ating and, wherever he stayed, usually a mountain resulted. And when he scraped the ground it became a salt lake. And wherever he stopped*

and regurgitated he produced a whole lot of ngin̠takas. *And what he brought up is now the various foods that they get in this area* (kal̠tu kal̠tu). *(*Kal̠tu kal̠tu *is one of the best seeds that they grind to make a cake.) And when his stomach got rid of everything it was empty and it was a big cave and there's a big cave you can see there. So this whole area is called* ngin̠taka *country. [Here Nganyinytja lapses at the spectre of all the sites and Ron prompts, 'And they still kept on following him—*tjiwa mantjinu?*'].*
Yeah, and when they killed him they found their tjiwa, *they broke it up and they killed him and there's a great big mountain system that he established there.*
I have finished this story and we shall go and see it all.

~

Ka pir̠uku atintja tju̠ta mayiku kunakan̠ti ngapar̠i tjanma̠ta ulkapa
And what he vomited up is now all the food plants in this area.

Illness and creation, transformation; in the *ngin̠taka* story, the *ngin̠taka* vomits out all the plant foods that *an̠angu* survive on.

The *ngin̠taka* story covers a vast stretch of country and we only see a small portion of the songline, six sites in all. The first, a rock shelter where *wati ngin̠taka* lies down to rest and hides the *tjiwa* behind a rock ledge; the second, the salt lake, he makes by sweeping his tail; the third, where his beard protrudes from the side of the hill; the fourth, the regurgitation site where all the food plants are created; the fifth, the wild onion dreaming site where he scrapes the seeds of the wild onion over the hill; and the sixth, a series of cave chambers which represent his empty stomach. It is the regurgitation site which most stimulates my imagination.

In the desert we drive through the country which lay beyond Nganyinytja's doorstep into the red rock ranges where we stop at each of the sites. At the regurgitation site we are led into the quiet of a circle of rock hills and stop beside an array of what appear to be ceremonial rock arrangements. Nganyinytja, Tjilkiwa, and the other women all talk and laugh excitedly. Low on the side of the hill, there is what appears to be a cleared track outlined with a row of rocks which leads into the ceremonial rocks, to a distinctive large slab of yellowish grey lizard-skin rock. The yellow rock is marked with small grooved circles, like the skin of the *ngintaka*, and lying on the rock are two round hand-sized stones. The tension rises and we are arranged for the performance in a row along the cleared track leading to the *ngintaka* rock. This track is Lionel's entrance; he appears now as *wati ngintaka*, dancing along the track to the lizard skin rock.

Nganyinytja says that when *wati ngintaka* arrives at this place, he vomits out everything in his stomach, creating the rock and all the *mayi*, food plants, that are now here. The *ngintaka* rock becomes the *tjiwa* and Nganyinytja shows us the two grinding stones sitting on its surface.

Tjilkiwa takes these two grinding stones and grinds the surface of the rock, painting her dark arms with bright yellow ochre, and begins to dance and sing. It is more powerful, she says, if the women dance with bare breasts here. I ask Nganyinytja whether it is a secret women's site and she says, circumnavigating my question:

minymaku watiku tjitjiku
it is for women, men, and children

We are safe, she is telling me, we are protected from sacred knowledge and are sharing what everyone can know; it is a public performance.

Ron translates the story of the vomiting site:

They're singing about how his stomach gave him trouble and he's about to bring up all sorts of food.
This is like the back of the *ngintaka*, these marks represent the seeds that he brought up.
And he ground up the seeds and peopled this land here with the plants that they eat.
Before the rains came he put out all the *uninypa*, seeds.

He ground them up and these too [rocks], he vomited them up.
They leave them there so they can teach their children about it.

Nganyinytja: *Yuwo-o!*

They grind food on that and that is an increase ceremony for plants to grow up in this area.
Just the same way as we keep seeds and put them in the ground, they do that, to produce the upgrowth of those two things, *kunaka<u>nt</u>i* and *ka<u>lt</u>u-ka<u>lt</u>u*.
They grind them wet, and it makes a sort of wet, floury mixture and they make the cakes out of it.
This is just ordinary rock but this is the dreamtime.

Nganyinytja: *Uwa. Inma tju<u>t</u>a kuwaripa* (now we will all sing)

Manta kantu<u>r</u>a nyaapa patapata<u>n</u>u
Earth stamping—I vomited out.
Stamping strongly on the ground, I project out in vomit all this food.

This simple song text comes to represent all that occurs in the confluence of performance and place at this site. In its translation I understand Benjamin's quote about the untranslatability of all sacred text and its absolute translatability. I understand the difficulty of the Two Women Dreaming search and at the same time I remember that my canvas envelope, my home in the desert, is *ka<u>lt</u>u ka<u>lt</u>u*. At this site I am *ka<u>lt</u>u ka<u>lt</u>u*, the seed ground by the grindstone and created and recreated in the *ngi<u>nt</u>aka* story. The remarkable confluence and concentration of meaning can be seen in the focus on the *ngi<u>nt</u>aka* rock, the symbolic focus for the increase ceremony for the *mayi*, all the plant food of the area. The rock, grey-yellow with a surface like wrinkled skin, was vomited up by the *ngi<u>nt</u>aka* man. The markings on the surface are the skin markings of the *ngi<u>nt</u>aka* itself; they also represent the seeds that he vomited up. The rock is also the *tjiwa*, the grindstone that the *ngi<u>nt</u>aka* hid in his tail, and so represents the original grindstone that ground the seeds of the plants he vomited to make them into flour; the rock is also the source of yellow ochre ground to make the body designs that represent the seeds/*ngintaka*/grinding stone. There is a multi-layered complexity of interlocking elements of performance and the presence of any one of these elements will powerfully evoke the others. In the absence of the site in the singing and dancing back at the camp, the place is as present as if it is here, evoked by song text, rhythm, melody, dance steps and body designs; the *ngintaka* and plant foods are created and recreated in a continuous present.

According to ethnomusicologist, Helen Payne (1989: 44–5), women's rites 'constitute a series of texts or sites' and sacred sites are 'those environmental features which bear constant testimony to the creative powers of the ancestors whose powers first shaped and now lie entrapped within them'. It is these powers, according to both Ellis and Barwick (1989) and to Payne, that are accessed through song. Payne goes on to discuss in a more generalised way the (sitemic) relationship between 'rites and sites':

> analytically speaking a site constitutes a topographic feature, often one of distinction ... all Dreaming sites constituted a waterhole and/or a vantage point from which to view surrounding countryside and/or a shelter ... the only sites to have such characteristics within an otherwise arid desert terrain (Payne 1989: 45).

and between sites and songline:

> From observing the colours on a landsat satellite image ... I found that as the sitemic relationship increased in the Dreaming, so too did the fertility of the region. Moreover, I found that the Dreaming path followed the only major line of fertility in an otherwise arid environment (Payne 1989: 45–6).

~

Inevitably we leave the desert and return to Alice Springs. Already I grieve for the dancing. I wonder about the meaning of the *ngintaka* performance today. It is so strong, so important that I want it to last forever, but the increase ceremony is no longer needed, the people buy flour to eat now. Even at Angatja, in the centre of the land, there are so few people who even know the songs to sing for the country, let alone actively perform them. Are they now simply performed for the sake of the 'tourists'? Nganyinytja is old and will not live forever; will the whole thing be forgotten? I begin to feel the pain of separation and loss, alienated in an environment where I can be nothing but tourist. I spend my time hovering in the multitude of shops that sell Aboriginal arts and crafts from remote communities. The smell, the feel of the places and the people is there, in the hairstring beads, the beautiful desert design fabrics and the now famous dot paintings. I go to the laundromat, buy tea at the pizza takeaway and walk back to Diana's across the silver sand of the Todd River, almost walking into a group of blacks drinking under the shelter of an old river gum. I circle widely to avoid disturbing them and one of the women, bottle in hand, recognising alienation when she sees it, calls out 'You're alright lady,' and gestures to me to come and sit with them.

The next day I hover back again in the direction of Aboriginal arts and crafts, like a moth to a flame. At the Araluen Centre there is the annual exhibition of art from all the remote communities in the region. I am stunned by what meets my eyes, my senses. Not the endless repetitive round of Western Desert dot painting imitations but the most alive, vigorous and exciting display of art I have seen. This hybridisation of culture, the production of artistic expression by remote Aboriginal communities for a hungry and diverse Western Market, has come of age. In one corner, a handful of paintings from the men at Papunya, the original location of Western Desert art, has matured into a distinctive and powerful simplicity; the dots are gone and strong, straight bare lines gouge the canvas in abstract landscapes of sombre traditional colours—black, white, yellow and red ochres. There are rows of silks gaily batiked with patterns from the desert—bush banana, wild plum, wild raisin—fine tendrils of plant and fruit designs. There are sneakers, earrings, note holders, boomerangs and all sorts of kitsch wonderfully decorated with bright desert dots or Namatjira-inspired realism. There is work from the women I have just left at Angatja, *iyininti* beads from Tjilkiwa and carved wooden animals from Nganyinytja. There is a host of beautiful dot paintings with design and text but one stands out among them all. There, amidst the red, yellow, black and white traditional colours is the surprise of a brilliant blue waterhole in a canvas of ironstone brown. I am back at Nganyinytja's waterhole and her front doorstep.

> Here third [world] is not merely a derivative of first and second, it is a space of its own. Such a space allows for the emergence of new subjectivities that resist letting themselves be settled in a movement across first and second. Third is therefore formed by the process of hybridisation which, rather than simply adding a here to the there, gives rise to an elsewhere within here or within there (Trinh 1993: 9).

There is such innovation, vigour, and adaptability in the representation of place. There has always been a tension between permanence and change. How can it be seen as a problem that the performance of desert songs has changed?

I return to Diana's, ready now to leave. She says when I get in, 'You might be interested in the text of a painting I collected today for an American exhibition—it's by Bessie Liddle, and it's about the Two Women Dreaming from Alice Springs. If you type it on to the computer for me you can take a copy.' I begin to type, feeling by now the weakness of the flu; my hands and neck ache, my fingers are leaden and my body seems unable to establish any relationship to the computer. Or perhaps it is the beginning of what I know will be a long haul.

~

Kungka Kutjara (*Two Women*)

The story coming from west of Amata (name Angatja/Pukara area) through Titjanardi; from there this painting takes the story. The women are travelling and eating bush tucker. The footprints follow their trails as they move from place to place. The story travels through Lake Amadeus towards Ulpanali *through the ranges heading for Alice Springs (two tits Arrente name). One man following is called* Arabie. *At* Ulpanali *(poison corroboree) he turned into a big rock at* Upanali *and the women went on alone.*

The women come through the Gap and travel along the Todd River. The Gap is a very significant site for this story. At the Gap an 'event happened' and the story becomes a part of the Arrente women's dreaming. The two women pass to the Telegraph Station (Arrente name I don't know). Here they were assaulted and managed to travel to the Alice Springs Dam site. Both were very ill. Here the Arrente Kungka Kutjara *stayed and this has become a sacred dreaming site for all women.*

~

Back in Armidale I walk in rain on a spring morning, sky uniformly grey, in another dot painting of pink petals on wet path. I smell the acacia and remember digging for *maku* under sweet smelling acacias in the desert. Eileen and I, sitting together under the *ilykuwara* tree, I am her apprentice. Tap-tap on the ground, I listen for the hollow sound and then the digging of the roots. Sometimes Eileen says, '*Tjaa, tjaa, anu*', as I pull out a swollen root with a hole and only a pupa case left; the grub has become a moth and flown away leaving only *tjaa*, its old skin, behind. Sometimes, much to Eileen's pleasure, there, in the wood of the root, lies a huge white witchetty grub, bursting, pushing at the boundaries of its tight inside home. Ripping the fibrous bark in strips with her teeth so as not to squash this delicacy, she lifts it gently from its home and lays it in my hand, its skin cool and moist, curled against the light. We eat them, skin just crisped by the hot sand beside a fire.

There are many Two Women Dreamings.

Performance III

Emily and the Queen

Landmarks
April, 1990
At the back a rocky outcrop stretches into a wisp of cloud, white against the bluest of blue skies. Air, shimmering blue from heat and eucalypts, carries strong blue eucalypt smell to the top of shiny granite rocks. In-between the two rocky outcrops lies a shallow arena about twelve metres wide clad in soft bleached-blond native grasses. Emily gets out and bends down, poking with her stick into the long golden grass to feel the elliptical scar of the graves marked by an outline of small rocks. After months of difficult negotiations and an arduous four-wheel drive trip up the mountain, we have finally arrived at the burial ground of the 'old Queen'. We are on the top of a mountain at Mooki on a privately owned station on a plateau of the Great Dividing Range in Northern NSW, out of a little tin mining town called Tingha. We stay in this special place for only a little while, not even long enough to boil a billy, and then bump and wind our way back down the mountain to picnic by the creek where Emily and her sister Marie remember their childhood.

In this performance I explore the interleaving of Emily's story with my own, and what meaning the visit to the old Queen's burial site can have for each of us. To understand Emily's performance I return imaginatively to this space of the Queen where I engage in a complex dance of meaning with Emily, and theorists Trinh Minh-ha, Turner and Carter.

Emily O'Connor and sister Marie Strong, Mooki, 1993
(Margaret Somerville)

Paul Carter, in his analysis of the new settlers' journals, develops the idea of a *space in-between*, opened up by performances such as those of Flinders' military personnel and the subsequent performance response from the 'natives'. Carter believes that what is important is not how the settlers and natives understood or misunderstood each other but the space of possibilities for symbolic exchange where maximum differences can be preserved and new movements can be devised between them. In my terms such a space-in-between remains a possibility today, an important space for Emily and me, a space for postcolonial transformations of meaning.

~

> This is not a moment of my life. It is a piece of my thigh. I've cut it into slices, between which I've spread out a few faded rugs. This is not my story. It's an enclosure and I'm unbuttoning my face there (Xavier Gauthier cited by Trinh 1991: 136).

Emily's mountain rises tall above the old homestead on a property called Mooki between Tingha and Bundarra, about eighty kilometres from where I now sit. But I cannot re-visit it; it is fenced (in, out?) by layers of fences, on private property, three properties in from the public road. And I have heard that the landowners are busy denying that there is a burial site there at all; they don't allow visitors. Emily and I, friends for just four years until she died in 1992, can no longer sit and talk together.

I rehearse the scene over and over again standing with Emily at the top of her mountain. I stand in shimmering heat rising above long soft grasses and follow the line of sheer granite walls reaching into blue. In front, my eyes stretch to eucalyptus haze hills through strange grass trees with tall brown stalks. I recall Emily's exact movements in that place, poking her stick through the grass to find the small round stones that mark each grave, the tone of voice and position of body as she speaks the words that surprise me so. *The graves have not been swept clean.* I am now sitting on top of the mountain; I become the mountain itself.

I visit that space over and over and know there is a profound connection between Emily's performance on top of the mountain and my ability to perform myself at this point; to make sense of my bodily experience in space, to story it for myself and at the same time for you, my reader.

> The liminal period is that time and space betwixt and between one context of meaning and action and another. It is when the initiand is neither what he has been nor what he will be (Turner 1982: 113).

6th December, 1993

A no-place illness. Stray blowfly buzzes round and round in still heat and the clock takes hours to move minutes. Can't eat, can't sleep, everywhere is toxic; nowhere to be. At last 2.30 and something to do. Inside, the Counselling Service reeks of paint fumes; I walk about outside, waiting. At three I go into a tiny enclosed room smelling of paint with a curtainless aluminium window making a blinding square of hot white light in front of me. I force myself to sit and talk. The counsellor suggests we do a relaxation. Overwhelmed by glare, heat, paint fumes and the presence of this strange man, I shake my head. 'You want to hang on to your fear forever?,' he says. 'All right,' I say, 'give it a go.' He brings in a yoga mat and I lie in shavasana, *the position of the corpse.*

I listen to his voice and sink deeper and deeper into a safe place within, the resonance of the voice cushioning the fall. The voice takes me into a beautiful house, a place of healing, and I realise I am at Warrawong;[1] *I try to displace it with other images but it will not go. A great sadness wells to the surface but it cannot come out, not yet. The voice brings me back to the room I am in, for this moment, still and calm. Where is the safe place?*

---

[1] Warrawong, the property where I grew up my four children and where I lived when I went with Emily to the mountain.

> Liminality may be for many the acme of insecurity, the breakthrough of chaos into cosmos, of disorder into order ... Liminality may be the scene of disease, despair, death, suicide, the breakdown without compensatory replacement of normative, well-defined social ties and bonds. It may be *anomie*, alienation, *angst*, the three fatal alpha sisters (Turner 1982: 46).

Bronwyn lends me *Journey through Menopause* to read which describes the disjunctive and often painful experience of menopause as a process of transformation. Although not menopausal, it allows me to re-define my state of dis-ease as a movement from one state of being to another, a liminal phase, but still I wonder 'But will I ever be strong enough again to open my mouth and not have a day of raw pain leap out?' (Lorde 1996: 61).

> rather than talking about death, I would prefer to talk about threshold, frontier, limit, exhaustion, and suspension: about void as the very space for an infinite number of possibilities; about the work effected on one side and the other of the limit, refusing to settle on any reductive position outside or inside, and instead making possible the undoing, redoing and modifying of this very limit. The work is brought to the borderlines, to a certain exhaustion of meaning, thereby suspending its closure (Trinh 1993).

I write a little, fragments, to make meaning of this experience. I am troubled by writing through this pain, I would rather conceal it from view. I read Susan Griffin and know that it is a matter of survival, that *every woman who writes is a survivor*, but

> I come back to this problem of despair in writing, myself caught up in it today, feeling a dullness about all language ... I feel as if my sleep had been disturbed, as if a dream were intruded upon, and I am not quite certain how to proceed. This is a profound disorientation. When I am not giving forth words, I am not certain any longer who I am. But it is not like an adolescent searching for an identity; no, this state of mind has an entirely different quality, because in it there is a feeling of loss, as if my

old identity, which had worked so well, which seemed to be the whole structure of the universe, were now slipping away (Griffin 1982: 230–31).

I met Emily in 1988 when Patsy and I were collecting stories for *Ingelba*. The three of us, and Emily's sister Marie, went to Ingelba and sat by the river while Emily happily told stories of her younger days visiting her father's family at Ingelba in a horse and sulky. It was not the stories that stayed with me however, but a promise that doggedly followed me for years. It came about like this.

Although Emily was happy to tell stories of her father's family for our book, she made it clear that she was writing a book of her own. Emily lived in the small town of Guyra in an ordinary housing commission house in a typical rural working class fashion. Her husband made money from 'digging spuds' in the season and the rest of the time they lived frugally on the pension. Neither of them knew any tribal language or any so-called traditional Aboriginal practices but, in the telling of her life, Emily defined herself through her mother's line and in the places of her mother's country, 'a flat plains girl' from west of Tingha. She traced her authority back to Maryanne Sullivan, 'the Queen of the tribe', whom her mother had lived with on the banks of Stoney Creek in the midst of pastoral properties.

It was on that day at Ingelba that she asked me to arrange a visit to the old Queen's grave, high on a mountain at Mooki near Stoney Creek.

It should be easy to write this performance, one woman, one place, one moment of performance; a place to crystallise this idea, to give it form. And yet it has taken so long. Five months I have been on the top of this mountain and I have slipped and tumbled, whirled back down to the bottom.

25th February, 1994

'A safe place,' she said this morning in the women's group, 'I am looking for a safe place.' Only the second meeting and a brave new

beginning, after our loss of women's studies. Bronwen is doing a Masters in Law and I've never met her before but I know exactly what she means. A qualified lawyer, married, with five kids, and still looking for a safe place.

> I think we're completely crushed, especially in places like universities, by the highly repressive operations of metalanguage, the operation that sees to it that the moment women open their mouths ... they are immediately asked in whose name and from what theoretical standpoint they are speaking, who is their master and where they are coming from (Cixous cited by Trinh 1991: 20–21).

After our day at Ingelba, I visited Emily every month or so in her tiny grey box-like housing commission house on the edge of the little town of Guyra. It was in her front room that I sat with her, curtains drawn, in the smoky haze of the old wood stove, piecing together stories of her life. I would bring a packet of shortbread biscuits and Emily would pour us cups of strong milky tea. After we had finished our morning tea she would get out her folders of writing, business papers and old photographs, and tell me where she was up to with her own project. Emily jealously guarded her writing from anyone's intrusion, and refused help, even though her hands were gnarled with arthritis and handwriting was difficult these days. She would show me what she had written since I had seen her last and then tell stories about her younger days. When I asked her about the Queen, she would tell me just a little:

> Mother told us how she looked after the Mission there and she was the Queen of the Mission—kept everyone in their place and didn't approve of strangers going there—and she was very strict. When the New England boys used to come down there poking around the Mission, she didn't approve of it, so when Mum and Dad went off together they had to go away, sneak away (laughs), so they went over to this other mission, Nucoorilma. They had to stay over there at Nucoorilma because they didn't want to come back and face Granny Sullivan. They come back after Bob was born.

14th March, 1994

Rituals of loss. I seem to have spent weeks tidying up Warrawong; ten years of family life to dismantle. It is two years since I've been here and the smell of despair hits me as I walk through the door. I wander through the chaos, not knowing where to begin. Then I proceed from room to room, cleaning as I used to every Saturday morning. My days develop a rhythm of sorting, cleaning, clearing. In the first room I place his hat and a beautiful handspun vest, his old tweed jacket, backpack and walking boots, clothes and linen, folded and stored in garbage bags; in the second, the double bed, the handpainted Italian dinner set and Anna's white cedar dressing table are ready to go to the kids; in the third room, there are the books, books of past lives—gardening, spinning, knitting, wool dyeing, preserving and jams, kids' books, books about parenting and building shelters, novels, poems and fairy tales—packed away in boxes and some given to the family next door. And then there are the papers, maps, birth certificates, qualifications, postcards and old letters—one, the last letter from my mother, written just weeks before she died in 1982:

> It's just beautiful sitting here in the second bedroom on a blue, sunny, cool afternoon, looking out across the park to the bay where all the little white boats are riding at anchor and pointing to the north east where the breeze is coming from ... They have decided against chemotherapy because they would have to give me such massive doses to have any effect on this particular type of tumour and it would make life a misery for me ... I just looked at the clock and it's six o'clock, so I had better get the tea ready.

I can't remember reading it at the time of my mother's illness; now I seem to be grieving for the loss of it all.

Finally the spirits of the place are settled; his things are packed in the blue truck; the kids' things are sent off in a removal van and everything else is organised for the clearance sale. I close the front door, put the key on the light box, and pick a bunch of bitter-sweet purple lavender which seems to be doing so well this year. I place it on the table in my new house.

One time when I asked Emily about the Queen, she gestured towards a photograph which stood out from the others displayed on the walls of her front room where we sat. This one was larger and mounted in a brightly coloured handmade red-checked gingham frame. Underneath the photograph was the inscription 'Queen Maryanne Sullivan'. There in the midst of a large group of people was a slight, dark, stern looking woman with a crown of some sort on her head and a band slung across her chest showing her name, M.A. SULLIVAN. I peered at the photograph taken at Stoney Creek Mission as Emily identified her mother and many of the others who were strangely captured in that image. There were about forty people there on that picnic day in 1909. They were all dressed in formal Victorian clothing; the women in high-necked blouses and skirts down to their ankles and the men in suits, vests and ties. The women's picnic hats lay on the ground in front of them, offerings to the camera.

I studied the face of the Queen. Who had taken this photograph and who had organised the picnic day? Who had vested Maryanne Sullivan with a crown and a name band and what did her queenship mean? What was this performance space arrested so temptingly by the camera in 1909?

> Contact history is spatial history. The meaningless gestures and sounds that criss-cross the meeting space, briefly bringing strangers into contact, serve to delineate a symbolic zone, to inscribe a space historically. The exchanges, whether of sounds, gestures, or objects, are like surveyors' lines connecting horizon points: their thickening network of sightlines, orienting this person to that person, this face to that face, help to map a common space, to characterise its behavioural and symbolic topography (Carter 1992: 179).

I questioned Emily further about Maryanne Sullivan. Why was she the Queen? Why had she been so important to Emily, and to Florrie (they were both writing about the Queen)? There were no further answers, no more stories were forthcoming. Florrie had by this time died and there seemed to be no one else who knew. I wondered whether Emily was the last in a long line of women who had passed on their stories to the next generation. As I left, Emily gave me a copy of the photograph and repeated her one request made some years earlier, for me to organise a visit to Queen Maryanne Sullivan's burial site.

28th March, 1994

I take down all the photographs from my notice board; wild women at Pine Gap Women's Peace Camp; Patsy and I at Ingelba; Marie, May, Maureen and Janet, the four women from Burrabeedee, and Queen Maryanne Sullivan's picnic day. All these women have kept me company during this long journey but I have failed in my negotiations with Patrick, my boss, about giving up my writing space, my womb room. He offers me a bigger room, upstairs in the main building, definitely an improvement, he says, on my little space at the end of a winding rabbit warren deep in the basement of the building next door. He wants all staff to be accommodated in the main building and thinks his offer too good to refuse. But I have come to love my exile, I think of all my photographs, my illicit conversations with women's studies students and friends, so necessary to my work, and I am reluctant to enter the world of Men and Language. I point out to Patrick that I am currently on leave and therefore entitled to stay in my student room. He is flustered, 'Office space and car parking,' he says, 'are the two most emotional issues on campus.' Not that we as academics ever write about space. It is just the ground on which we build our sense of ourselves in this institution. I remind Patrick that I was originally housed in the main building but a professor decided that he should have a room three times as big as mine so asked me to move to my little space in the building next door. Patrick says that was in the past, water under the bridge, and I felt suitably ashamed for airing the dirty linen in public, such things are not to be spoken.

> *we are left with the old definition of dirt as matter out of place. This is a very suggestive approach. It implies two conditions: a set of ordered*

relations and a contravention of that order ... Dirt is the by-product of a systematic ordering and classification of matter, in so far as ordering involves rejecting inappropriate elements (Douglas 1966: 35).

I agree to move, but where is it safe to do this writing? I keep wanting to erase what I have written so far, that sense of being on the borderlines, somehow shameful and polluting.

Emily's performance was cancelled so many times. The landowners initially refused us permission until I persuaded them that Emily was a very old woman and it was her one last wish to visit the burial site of the old Queen. I said she wasn't about to invade their property, in fact she could barely walk and then only with a walking stick. We would have to drive her right up to the site. They said it would be very difficult to get there and we would need a four-wheel drive. I rang Emily. 'Can you remember how easy it was to drive up the mountain last time you went there?' 'Oh Margaret,' she said 'last time we went there, it was in a horse and sulky.' That was forty years ago. I organised to hire a four-wheel drive vehicle but each time we were due to go the clouds opened up. It rained and rained and the creeks swelled and burst, cutting off our access to the mountain. After two weeks or so of clear weather, on the 7th April, 1990 we decided to have another try. In the morning I woke to a leaden sky and even my son said maybe the spirits didn't want us to go there. This time I was determined that it was now or never, so I set off and on the way to Guyra I was relieved to see a small patch of blue. My mother had always said if there's enough blue to make a sailor a pair of pants ... I picked up Emily and her sister Marie in best picnic clothes with billy cans and picnic bags and by the time we turned west to Tingha the little patch had opened to a wide intense blue.

5th April, 1994
I decide reluctantly to spend Easter at home to settle myself for this writing and am reminded of Trinh's story about Florence

Edenshaw, a famous American journalist who, when asked by a woman what to do about her self concept said 'Get dressed and stay home'. And Trinh's comment:

> *Edenshaw's statement remains multilevel, it ultimately opens the door to a notion of self and home that invites the outside in, implies expansion through retreat and is no more a movement inward than a movement outward, towards others (Trinh 1993: 5).*

I have now spent three days in a frenzy of tidying and cleaning. The house is tidy and there are clean sheets smelling of sun; the linen chest is repacked with summer quilts stored and enough linen in case the kids visit; the back room is emptied of leftover boxes and rubbish; the vegie garden is extended revealing its newly turned, lush chocolate-brown earth and the grass is mown and clipped.

I watch the thoughts come and go as I methodically pull out layer upon layer of matted white couch roots from the mulch in the drive.

Office, garden, house and mind. Am I now ready to arrive at this place of performance?

After crossing the cobbled black riverstones of Stony Creek we headed towards the last mountain. The track ended just past the homestead and we slithered our way across a huge wet area at the base of the mountain where spring water seeps across flat rock covered by moss and soft green sedge grass. The wheels slipped and slid as Emily directed none too carefully with her walking stick out of the car window. The ground became drier as we wound our way higher through a maze of lichen-covered boulders and ever-thickening eucalypts. After what seemed like hours of tedious winding and bumping with Emily cheerily re-membering from the back we arrived at a protected dip in between two spurs of the mountain.

At the back of the dip a tall rocky outcrop stretched into a wisp of cloud against the bluest of blue skies. In the front a circle of boulders surveyed the ring of distant blue mountains. Even the air shimmered blue from the eucalypts, hot in the reflected heat of the rocks. In-between the two outcrops lay a shallow arena about twelve metres wide, clad in bleached gold of soft native grasses.

8th April, 1994

9am: Already there have been three calls about the clearance sale and one of the goats has picked today to die. As I walk, a myriad of thoughts pass through my mind and blow away like the wind. I think of the third eye, the mind's eye and Emily and the power of visualisation/imagination and a lot about the safe place. That is what has resolved the dilemma of continuing with this writing, no longer tracking only the course of pain but the building up from that most basic and fundamental building block, the safe place inside.

1pm: As the clearance sale starts I light the fire in my new house and burn, with walnut twigs from the back garden, angry letters sent to me over the last two years. The fire crackles and sparks until only the finest white powdery ash remains. I light incense in every room and friends arrive, unexpectedly, to share the late afternoon sun.

Where is the safe place when all else is let go? Not Warrawong, it is gone, not my office, it is gone, not my little house ...

> Walking on masterless and ownerless land is living always anew the exile's condition; which is here not quite an imposition, nor a choice, but a necessity. *You'll learn that in this house it's hard to be a stranger. You'll also learn that it's not easy to stop being one. If you miss your country, every day you'll find more reasons to miss it. But if you manage to forget it and begin to love your new place, you'll be sent home, and then uprooted once more, you'll begin a new exile (Blanchot).* The work space and the space of creation is where she confronts and leaves off at the same time a world of named nooks and corners, of street signs and traffic regulations, of beaten paths and multiple masks, of constant intermeshing with other bodies', that are also her own—needs, assumptions, prejudices and limits *(Trinh 1991: 26).*

I move towards letting go.

The fifteen or so graves were hidden in the bed of golden grass. Emily got out of the car and I helped her through the long grass and stony ground to the graves and stood back a little to give her space. She felt the elliptical scar in the trunk of the old yellow box tree where the bark had been removed for a burial. She was contemplative as she poked around the graves with her walking stick, discerning the position of each one by the arrangement of small rocks outlining their shape. Then, suddenly and peremptorily, she looked up and she spoke:

The graves have not been swept clean.
Just that, no more performance, no adulation, no secrets from the place itself.

> How, then, are such spatial events to be represented, introduced into history and made part of our contemporary consciousness, without destroying their poetic self-sufficiency? ... The appropriate way to represent the characteristic events of contact history is theatrically (Carter 1992a: 182).

Thursday, 12th April, 1994
My whole body taken up with this writing, patterns and thoughts flying through my mind. Spent yesterday afternoon crying to Lillian about my profound sense of loss for Warrawong. Afterwards I walked out at Dumaresq Dam alone, late afternoon, glad of the company of the dog and the sound of waterbirds on that body of water. That night, tucked up in bed, I read Paul Carter's Living in a New Country *and a great excitement wells up from my belly, of connection, recognition. His chapter about performance and history—different from my ideas about performance but exciting intersections—I skim read so as not to engage too deeply with ideas that will keep sleep away. As I fall asleep, I wonder about this male writing that I admire so much, why I can't write in this persuasive voice, the accomplished ease and wholeness of the word, the erotic power of persuasive language. And Victor Turner too, whose concerns are always universalizing; the compelling power of grand theory. I admire and desire their language of persuasion, soaring towards the angels, while I scrabble on the ground amongst the fragments. But I am also frustrated that there is no body presence*

in their writings. From whence come their obsessions; who are they? And another voice in me says I don't write like that because I choose otherwise.

> 'Writing the body' is that abstract-concrete, personal-political realm of excess not fully contained by writing's unifying structural forces. Its physicality (vocality, tactility, touch, resonance), or edging and margin, exceeds the rationalized 'clarity' of communicative structures and cannot be fully explained by any analysis. It is a way of making theory in gender, of making of theory a politics of everyday life, thereby re-writing the ethnic female subject as site of differences ... Woman as subject can only redefine while being defined by language (Trinh 1989: 44).

So many ideas, feelings, coming all at once. I spoke to Paul Carter thus, wanting to write him a letter, make a connection. I will read his chapter carefully today. It is as if the drought has broken. I have sat with Emily's performance and Victor Turner for so long it has become inturned and would not be born. Now Paul Carter has joined us, and Trinh, always there, has made her presence felt. The waters have broken and there is a strong sense of physical becoming ... perhaps this chapter will get written just like this.

It is this moment of performance that I have held in my mind's eye and understand now, for the first time, the dilemma of representing it. With *Ingelba* and *The Sun Dancin'*, I could return to the place with whoever was present at the time, including all the main actors, and then write the script of the performance as it happened. This has been a necessary part of the process of representation of landscape stories when moving back and forth between self and other across a space of multiple difference. In this case, however, Emily has died and there is no possibility of return to the actual place.

But I remember the complex association of place, story and performance in the desert with Nganyinytja:

> Rites ... involve the simultaneous presentation of a multiplicity of phenomena, viz musical forms, dance steps, and actions, painted designs, ceremonial objects, the naming of topographic features and/or of persons and so on. For the enculturated participant each phenomena is mnemonically associated with the

others so that all are to some degree simultaneously present at any single moment of a ritual performance (Payne 1988: 13).

Perhaps I can call up the moment of performance by re/membering Emily's words and gestures, by calling up the place. And maybe Emily does her own writing in this return to the space of the Queen. So I return to this place of performance in my imagination and pause with Emily in this space of the Queen. This time we have with us Paul Carter, Victor Turner, Trinh Minh-ha, in a different performance, a meeting place of ideas, a dance of intertextuality.

> Its discursive model is not the monologue but the dialogue: the distance, the space between the self and interlocutor, is ... the indispensible ground of two-way communication. But for the existence of the other, there would be nothing to do or say, no event, no song and dance (Carter 1992a: 176).

I have come to know Turner's ideas well, not only from his own writings, but through Richard Schechner, Turner's most committed disciple, writing from his position as director/theorist of experimental theatre in New York.

Turner had a lifelong fascination with the Ndembu tribe of Africa where he spent much of his early field work. On his return to North America he translated his understandings of Ndembu rituals to his own urban social context. It was ultimately the relationship between the rituals he studied in Africa, and the parallel *social dramas* that he found in his own culture, that occupied the rest of Victor Turner's imaginative and intellectual life and it was in this that his work resonated with that of Schechner in experimental theatre:

> Turner sought to integrate the notion of liminality—the threshold, the betwixt and between ... with his emerging understanding of the relationship between social drama and aesthetic drama. Performance is central to Turner's thinking because the performative genres are living examples of ritual in/as action (Schechner 1986: 7).

Turner's major focus in these social dramas, and the source of intersection with my own obsessions, is the *liminal* phase, a term he borrowed from anthropologist Van Gennep's *limen* or *threshold*, meaning *the space of becoming* for the initiand in initiation rituals. Schechner describes Turner's obsession with the liminal as living in a 'house that was all doors':

> Turner, who specialised in the liminal, the in-between, lived in a house that was all doors: every idea led to new ideas, every proposition was a network of possibilities. I think he was so long interested in performance—theatre, dance, music, ritual and social drama—because performance is the art that is open, unfinished, decentered, liminal. Performance is a paradigm of process (Schechner 1986: 8).

It is Paul Carter's work about first contact meetings between new settlers and Aborigines that draws me in. In *Living in a New Country* he analyses descriptions from journals and diaries written by first settlers, and develops the idea that what is happening in these events is that a symbolic space of exchange is opened up between the two groups of people, a space in which differences are preserved, and meanings remain open. He explores such an exchange between Matthew Flinders and the Nyungar tribe of Western Australia in 1801:

> Our friends, the natives, continued to visit us; and the old man, with several others being at the tents this morning, I ordered the party of marines on shore, to be exercised in their presence ... when they saw these beautiful red-and-white men, with their bright muskets, drawn up in a line, they absolutely screamed with delight (Flinders cited by Carter 1992: 161).

After substantial detailed analysis of this event and its effects, Carter suggests:

> The meaning of the event ... consists in its ability to create, albeit briefly, a provisional mode of exchange, a physical and symbolic space inscribed with meaning. But the meaning lay in neither the future nor the past, but in the moment itself when two culturally diverse groups of people without a language in common made contact (Carter 1992a: 163).

While Turner focuses on the experience of the performer/s and the liminal state entered into by the initiand/s, it is unclear what part the audience plays, or who exactly constitutes the performance. Carter, on the other hand, focuses on the space in-between two groups or players who do not share a common understanding, and the space that opens up is the space between the two, a symbolic bora ground in which the two groups may participate in a dance of difference. The dance represents a movement across difference, 'a thickening network of sightlines' that 'help to map a common space'.

I realise, in my reading of Carter, that I see contact events as not confined to the past, but that each performance in the landscape with Aboriginal people has had for me these same characteristics of first contact events. In *Ingelba*, when we walked around the site of the old reserve mapping the relationships to hearths and landscape; in *The Sun Dancin'* when we sat on the ground telling stories in the cemetery; and now with 'Emily and the Queen', meanings are made on each occasion as if for the first time. For me it is a confirmation of new possiblilites for re-visioning this land, for telling new inclusive stories only made possible by the organic existence of these stories in the minds of the players which make such performances possible.

~

In this I saw exciting intersections, not only in the theorising of Emily's performance and other Aboriginal performances in this landscape, but with contemporary feminist theorising of the self and other which has attempted ways of understanding this movement across a space of difference in which one does not appropriate the other:

> I want to run the risk of speaking within the space between myself and another's self ... my transgressions are aimed at transforming the representation of the relations of alterity; my limit-attitude can be characterised as the rejection of a binary construction of identity as horizon (Probyn 1993b: 145).

I return again to the top of the mountain, my performance space, and move to the limit/edge of understanding. It is a lively and complicated dance with Trinh, Probyn, Carter and Turner dancing with Emily and me.

> In the context of representing primary spatial events, such a performance would not focus on the dancers but on the choreography of space as such (Carter 1992a: 184).

In the choreography of this performance the question is ultimately one of whether I can know how to do this dance.[2] I begin by visualising Emily's space of the Queen. Why did Emily return over and over, symbolically, to the idea of Maryanne Sullivan, Queen? What sort of safe place did it represent for her?

~

The space of the Queen is above all a liminal space between different ways of being and knowing. Geographically, materially and symbolically the burial site is set up in opposition to the well established land use/spatial practices of pastoralists in the area known as the western slopes of the Great Dividing Range. Each property is signalled by its name, by boundary fences marking the lines of exclusion and ownership, by roads leading to station homesteads and other outbuildings, by cleared paddocks serviced by man-made dams and by the presence of sheep and cattle and other paraphernalia of the grazing enterprise. This enterprise is fully supported by government, legal, and financial structures and by the dominant stories of our culture. The metaphors of this spatial practice are of individual ownership, consumption and production. Multiple examples could

---

2 Both Cath Ellis and Helen Payne report incidents where they inherited traditional Aboriginal women's performances where it was literally, as well as metaphorically, a question of knowing how to do the dance. They both discuss at length the practical and theoretical challenges this raises.

be chosen from geography, agriculture, Lands Department and Mining Department maps and so on; it is a phenomenon with which we are so familiar that we take it as the natural state of being. Emily, however, never questioned either her right or ability to go to the top of the mountain and enter the space of the Queen, she simply insisted that I, as at least in part representative of these dominant cultural practices, should negotiate it.

The space is physically and materially located in a natural amphitheatre; high up on the mountain away from fences, farm buildings, improved and cleared pasture land. It is protected as well as exposed, a brief interval of shelter with a ring of high granite outcrops reaching into sky behind and a circle of granite boulders in front, a natural platform facing a range of blue hills in the near distance. It is this physical space in the landscape that I return to over and over in my imagination as I write.

In the centre of the enclosed circle the fifteen or so graves are organised in a particular relation to each other. There are two rows of elliptical shapes all facing the same direction, towards the tallest of the blue hills spanning the horizon to the north west. Although Emily no longer remembers which grave belongs to the Queen (*the graves have not been swept clean*), she says that the Bunawanjin Mountain was special to her and that the Queen was buried facing the mountain. The choreography of space signals the story. In this space there is a proliferation of different stories, alternative possibilities:

> Liminality can perhaps best be described as a fructile chaos, a storehouse of possibilities, not by any means a random assemblage but a striving after new forms and structure, a gestation process, a fetation of modes appropriate to and anticipating postliminal existence (Turner 1990: 12).

Because Emily had so few stories about the Queen I searched for a long time for other stories about Maryanne Sullivan.

Old Florrie Munro told me that when she was a young girl, she and the Queen's son had wanted to get married.

Maryanne had questioned her about her totem. 'What is your meat?,' she asked Florrie. Florrie, not knowledgeable in the ways of the people, replied 'Sheep.' Maryanne thereby decided that they should not marry because of Florrie's lack of traditional knowledge. Later, she was prevailed upon and declared that if they wanted to marry they would have to undergo a test of endurance in the bush. The young couple therefore set out on an arduous journey from Stannifer to Nucoorilma on foot, and without food and water. When they completed the journey they were allowed to marry and a relationship of great mutual respect developed between Florrie and the Queen. When Florrie told me this story at the age of eighty-eight, almost blind and crippled with arthritis, she was writing her story of the Queen.

Clariss Connors could still remember that Queen Maryanne was a firm believer in using the *nutta*, club, and was very experienced in using it to kill small game. She said Queen Maryanne died sometime in the 1920s when Clariss was a young girl of about fifteen. She said Queen Maryanne was buried on top of her two husbands, as this was the custom of the Barnbai tribe, and the body was wrapped in bark. She loved the mountains and was buried overlooking Bunawanjin Mountain, (Blackfellows' Waterhole), which is significant to the Aboriginal people of Tingha because of the story of the black wallaroos. After the tribe left Stoney Creek on Bassendean Station and moved south to Bundarra, the black wallaroos were said to have left Bunawanjin Mountain and never returned (National Parks & Wildlife Service 1976).

In a collection of stories from Tingha, called *Nucoorilma*, after the apple gums or angopheras, the people said that 'Granny Sullivan was a truly remarkable woman with a wonderful knowledge of the secret life of her people.' She was known by everyone as *meengha*, the mother of all mothers, and because of this she had three husbands. She wore a special kangaroo cloak. She arrived 'mysteriously' at Stoney Creek one day as a young woman, and may have been the daughter of King Billy of the Kamilaroi tribe who had three sons and seventeen daughters. One of the daughters, Mary Anne, went missing, possibly stolen by Grandfather John Munro who later became Maryanne's husband.

There were political stories too, such as the one about lost title deeds which resonates with many other Aboriginal stories about land:

> Mr Tom Lansborough tells us that Mr Wiseman gave back Stoney Creek to Granny Sullivan for the Aboriginal families. Some of the old people recall seeing a document to this effect but presume that when Granny Sullivan died, it, as with all her possessions, was burnt according to custom (Fennell and Grey 1974: 155).

There was another story of a different type, a representational story, hidden in a special back pocket of the book because it was much bigger than a normal page. The folded sheet, many times the size of the pages of the book, when opened out to its full size, takes up half my office floor. It is covered with rows of tiny print joined by myriads of lines and, at the top, two lone names: Maryanne Sullivan and John Munro. In *Ingelba*, Patsy Cohen represented the genealogical story of the five matriarchs as a series of concentric circles. Here the genealogy of Maryanne Sullivan is represented in terms of the more familiar written form of the family tree, but the names sprawl and grow in all directions so that it can never be fitted into A4 format, even with the print so small it can barely be read. Rarely more than five generations deep, these complex and multitudinous genealogies from oral cultures tell of lines of connection spread across the landscape. Among the hundreds of names, I find Florrie and Emily and see a visual representation of a much larger story: where they fit in relation to each other, to Maryanne Sullivan and to the rest of the people represented there. Through this story all the names are linked to Maryanne Sullivan and to this space of the Queen, at the same time constituting the story of the Queen as larger than an individual story. It is a story that spreads out laterally across generations, like Deleuze and Guattari's rhizome, or Foucault's *genealogy*, knowledge through surfaces, rather than depth.

I ask now, what can all these stories mean to me? How can I be made in this liminal space of the Queen? In what ways was it 'a safe place' for Emily, a space that she could move in towards and out from, a space from which she could write? Can this be a safe place for me?

> Exchange was not by way of symbolic objects—the biscuit might as easily have been beads or a looking glass—but through the creation of a mutually intelligible symbolic form (Carter 1992a: 173).

I move in my imagination between Emily's and my symbolic space of the Queen. Maryanne Sullivan made a new beginning; her story is about the re-negotiation of space after loss and dispossession. The story does not tell of Maryanne's origins in the place, it tells of how she came to the place, in a sense from nowhere, and began a new story. Emily's Queen had legitimate public power too. In her one story of the Queen, Maryanne had ultimate authority in determining how the people lived: 'She was very strict' ... Emily's mother and father had eloped to another Mission because the Queen did not approve of the liaison and they didn't return until Emily's eldest brother was born.

Perhaps the ideas encompassed in the word *queen* were the closest the people could come to their own word *meengha*, which signified notions of embodied female power, mother of mothers. The image of the Queen was also associated with past, barely remembered and no longer practised tribal customs; she wore a special kangaroo fur cloak (perhaps a sign of her status) and she knew how to use a *nutta* for hunting small animals; she assumed some sort of relationship to the legendary, probably totemic, story of the black wallaroo of Bunawanjin mountain. She was also a figure of mystery, possibly a daughter of King Billy with inherited royal status? Like the five matriarchs of Ingelba and Nganyinytja, Maryanne Sullivan straddled two eras of Aboriginal (his) story, the time before and the time after white settlement. She made it 'good for the people to go forward'.

All these women seem to have derived their status and authority precisely from that conjunction of femaleness, Aboriginality and the land through which the European settlers had inscribed the new land for themselves.[3] They were female, they were Aboriginal and they were the land, and for many extraordinary and ordinary reasons they inscribed their own story on their landscape.

The image of the Queen used to inscribe this space of symbolic exchange was that of Queen Victoria, because of the extensive use of Victoria as symbol of colonial power and authority during the time when these women were coming into power.[4] The reign of Queen Victoria parallelled the height of these women's lives, around the turn of the century. It is likely that she also performed a significant crossing over role, as with Nganyinytja, but Nganyinytja's later time frame lends itself to a different discourse.

My stories of the queen had also been different ones. I can remember being awed as a child by the impressive pomp and ceremony of the coronation of Queen Elizabeth II with her exquisitely jewelled crown and long dark red velvet train with ermine trim; and the excitement of standing on a box in a crowd lined several layers deep along the roadside in Sydney when the Queen visited in 1954. The power of the image was diluted gradually with continuing exposure to the royal family, in the *Women's Weekly*, with stories of corgis and holidays at Windsor Castle.

Then there was the final degradation of multimedia exposure to Princess Di and Prince Charles' marriage and separation which continued to absurdity with Fergie and

---

3 According to Holloway (1993), the tropes of women and Aborigines were used to inscribe land in the new colony.

4 Diane Barwick (1978: 51–64) has commented on this in her essay on the missions in Victoria; she associates the image of Queen Victoria with the rise of women's power on the missions.

Andrew. It seemed that, not only in the fairytales, princes and princesses steal the limelight, subverting a notion of women's power to the vagaries of the romantic love storyline. Besides, both fairytale and real queens were of European origin and seemed to have little to do with contemporary life in Australia.

Then this new image of the Queen began to appear in my work. Not only was I presented with insistent stories of Aboriginal queens in Maryanne Sullivan and Mary Jane Cain but Sidonie Smith, in her extensive work on women's life scripts and autobiographies, suggests that the queen is one of four life scripts within which women can claim authority in the public domain:

> The second female figure of legitimate public authority in the Renaissance was a political one: the queen. Several women who came to thrones during the Renaissance—Mary, Queen of Scots, Elizabeth 1, and Catherine de Medici—might look back to the powerful women of the feudal aristocracy as predecessors (Smith 1987: 34).

So the space of the queen came to represent the possibility of a place in the Australian landscape where there was an established authority for a woman to speak. Emily's space of the Queen was first and foremost a physical location that represented a liminal space for the exploration of alternative possibilities. It was a safe place that could be returned to in retreat and moved out from again into the world of public writing.

But here in this space of the Queen how did Emily choose to represent her performance?

The graves have not been swept clean.

Part of the shock of that moment was the surprise of the juxtaposition of the two metaphors, the metaphor of royalty and the metaphor of domesticity. If a single metaphor is a bridge, a way of moving from the known to the unknown, then the movement between two metaphors is a complex dance.

Sweeping, tidying, cleaning, mental cleansing ... In contemplating the metaphor of sweeping, I remember the wailing women in the central desert sweeping the ground around the camps when someone had died to erase the footprints which tied their spirits to the earth. I remember Maureen and Marie squatting to 'sweep' the graves as soon as we arrived at Burrabeedee Cemetery—weeding, digging, rearranging decorations—as they told stories of the people who had lived there; I remember Mother's Day at Burrabeedee when the orange-pink earth of the whole cemetery was swept clean and a woman who had travelled from Woollongong talked to me about her mother while sweeping her mother's grave; and the day spent with forty people clearing and cleaning Woolbrook Cemetery, erecting white wooden crosses to make the Aboriginal graves visible.

A highly complex image, this sweeping the graves clean; of spirits and memories, place and people, life and death. And each of these activities I associate with remembering, nurturing the memories of the person concerned and the network of interrelationships represented by the women involved. The physical material act in place was inseparable from the emotional, cultural, spiritual work that was being done.

And yet Emily's graves have not been swept clean, the Queen's grave is not visible. As she made this statement, Emily felt through the grass with her walking stick to find the pattern of fist-sized round stones outlining the elliptical shape of each grave. It was the *reading* of the signs in the landscape that was critical to Emily. Emily's metaphor is one of visibility/invisibility:

> My work there brought to my attention a singular and dominant theme, the people's severe invisibility and the consequent disturbing and psychological consequences of being unnoticed ... Often it is merely that such people are not seen; they are treated as invisible (Meyerhoff 1986: 263).

Emily's metaphor is not only about the in/visibility of the Queen's grave but the in/visibility of a whole set of spatial practices counterposed to hegemonic white spatial practices of land tenure and land usage. How can the people read themselves back from the landscape when the signs have been erased? If all such performances function as first contact meetings, then the meanings of this space are multiple and diverse, constituted across the differences between Emily and me, a space which remains open until there is a visibility of an entirely different set of spatial practices. In the context of her performance, Emily is making visible an alternative cultural approach to the relationship of people to this land:

> One of the most persistent but elusive ways that people make sense of themselves is to show themselves to themselves through multiple forms: by telling themselves stories; by dramatizing claims in rituals and other collective enactments; by rendering visible actual and desired truths about themselves and the significance of their existence in imaginative and performative productions (Meyerhoff 1986: 261).

I could dance this dance of meditation forever fluttering back and forth between the two metaphors, never able to pin the meaning down. In the end it seems that:

> The logic of this playful dialogue is not determined semantically: the aim is not to translate or interpret, not to find a common meaning—if the meaning of this dance could be written down, there would be no need to go on playing—but to find a system of communication where the greatest differences can be expressed simultaneously and, instead of cancelling each other out, be instantaneously transferred from one side to the other (Carter 1992a: 180).

I return again to the place of performance, the security of narrative, to stop the words from flying away.

Not knowing how to respond to Emily's words, I feel a strong injunction to clear the graves but I do not know how. Instead, Emily asks me to photograph herself and her sister Marie at the site.

Then we leave just as dramatically as we arrived, no lingering for the long awaited billy. Nothing further was said about the Queen, just that we should leave now and go back down the mountain to the site of the Stoney Creek Mission There, Emily makes herself comfortable and immediately begins to storytell her life into the tape recorder.

As we picnic at the spot where Maryanne's picnic photograph was taken in 1909, Emily tells stories of roaming through the country as a child. She remembers hunting parties; the ritual purification by smoke at the burial of her grandfather; a death and burial up on the mountain. By the time of her childhood the people had moved to Bassendean Station next door and she remembers collecting rations from the manager's store and the site of her birth there. Emily and Marie have never actually lived at the Stoney Creek site but they fossick eagerly along the banks of the creek where, out of reach of the plough, they find many relics of the campsite which they load into the truck to take home to display in their front gardens.

There seems to be a direct relationship between the space of the Queen and Emily's authority to story herself, and yet she refuses to tell her story in that space. It was to remain liminal, a space of paradox. It was not until we had completed the ritual performance in the space of the Queen and had left that space that Emily was willing to tell her stories.

~

Back in my office I transcribe tapes of the stories, organise photographs and make a book each for Emily and me with a photograph of Maryanne Sullivan on the front framed in a red-checked gingham cover, like the picnic photograph on her wall. In my book I display the photos and write the story of the day, hesitantly, in handwritten pencil, feeling a need to 'tread lightly' on this land. In Emily's, I set out the photographs and leave the spaces blank for her to write in her story but she wants me to write the story in her book. We talk a little about her writing, how she is finding it increasingly difficult to keep going with her hands bent with arthritis but she is resistant, as usual, to my help. Her resistance seems related to her sense of herself as writer, her strong need to retain ownership, authorship. We talk about how she might use the transcripts of the stories from Mooki but Emily is determined that it is the written form that she wants. She rejects her talk stories and I, on the other hand, believe the written autobiographical form is hegemonic. I wonder what the day has meant to her, whether she might include anything of the day in her book. My red-check gingham covered copy sits on my bookshelf, the story not at all resolved in my mind. I struggle with the narrative of the queen, the writings and the storytellings, the place and the lives intertwined, trying to make sense of my search. A leader, a model, a vision of possibilities, born from this earth, made out of this landscape, a new beginning?

I realise now the imperative for Emily: if you can't read the signs in the landscape, you must write them. It is the way these women have written themselves into the landscape, the public arena and can read themselves back. Not only this, but through the space of the Queen both Emily and Florrie found the legitimate authority to write themselves into a new public arena. She had provided a space of crossing over, of transferring authority from one realm to another. The image of the Queen is about voice, authority, text and visibility. It was also in this space of the Queen that Emily and I negotiated a complex relationship between orality and literacy. Emily, doggedly committed to the written form, and I, committed to a more direct movement across from the spoken word into writing. During Emily's life there appeared to be no resolution of this negotiation between orality and literacy. She never allowed me to assist with her writing or for the spoken word to appear in her text. The space between us was not about resolution but about movement and difference.

~

I don't see Emily for some time after I give her the red-checked book. I have left her with my phone number and the offer to help in whatever way she needs. I hope that in the intervening time she is making progress with her autobiography. But the people begin to despair that Emily's book will ever be produced. Then, one day in July, 1991, I have a phone call out of the blue. It is Emily and she says she has finished her book and she wants me to come and pick it up and prepare it for publication.

I drive up to Guyra only to find that Emily, instead of being at home as I had supposed, is in hospital. The family tell me to go to the hospital where I find her sitting up in bed, pink bedspread, pink bed jacket, translucent skin smiling too brightly. 'It's only arthritis,' she says. I try not to notice how diminished and frail she looks. Emily talks excitedly about the completion of her book and her detailed plans for the launch. She says she has one final task to complete. I avoid acknowledging what I already know, what we both know. 'I'll leave it with you,' I say 'and you can phone me when you've done the final bit.'

A phone call comes a week later from the hospital to say that Emily has lapsed into unconsciousness and is not expected to live. 'What is happening with the book?' the people ask. I sit talking with her for one last time. I talk to her about her book and how it will be published. Holding her hand, tissue paper skin stretched across bone. There is just the slightest sign of recognition, an almost imperceptible curling of her fingers around mine. I hesitate to ask it of her, wanting just to allow her to go with ease. Emily is buried in the Tingha cemetery and the people sing 'When We Meet on that Beautiful Shore' and mourn the loss of a leader. I don't know whether she is buried facing the mountain.

After it is over I take her manuscript back to my office.

~

Now nearly three years after Emily's death I realise how powerfully Emily's and my lives intersected in the space of the Queen. I worry about this writing I am doing about her but I know that in this ritual of re-membering I can do no less than to think through the dramatic issues her performance raised. I worry that I have overemphasised the idea too much, distorted Emily's meanings and I contemplate Emily's manuscript afresh as I return to it. In it, I discover, much to my surprise, a palimpsest of memories, layer upon layer about her two visits to the space of the Queen, one forty years previously with her mother, and the second, when we went there together:

> On our walk [in 1953], they took us up to the cemetery and it was all nice and clear then and you could see the graves. Our mother and the other elders showed us which grave it was and they told us that our queen was buried there. We did the graves up and we could see clearly where they were buried but when we went back the second time, just recently, we couldn't see a thing because of the long grass. That saddened me, not being able to see where our Queen was buried, our Grandmother, Granny Maryanne Sullivan (Connors manuscript).

In the space between orality and literacy, the marks of the Queen's grave are as important to Emily as they are to me. The figure of the Queen in the Australian landscape offers us both a reference point, a safe place to return to from which to write. It offers a beginning point out of the formlessness to shape the contours of the writing in response to the contours of the landscape. A place from which to imagine new myths and rituals of place to perform another Two Women Dreaming in this space in-between.

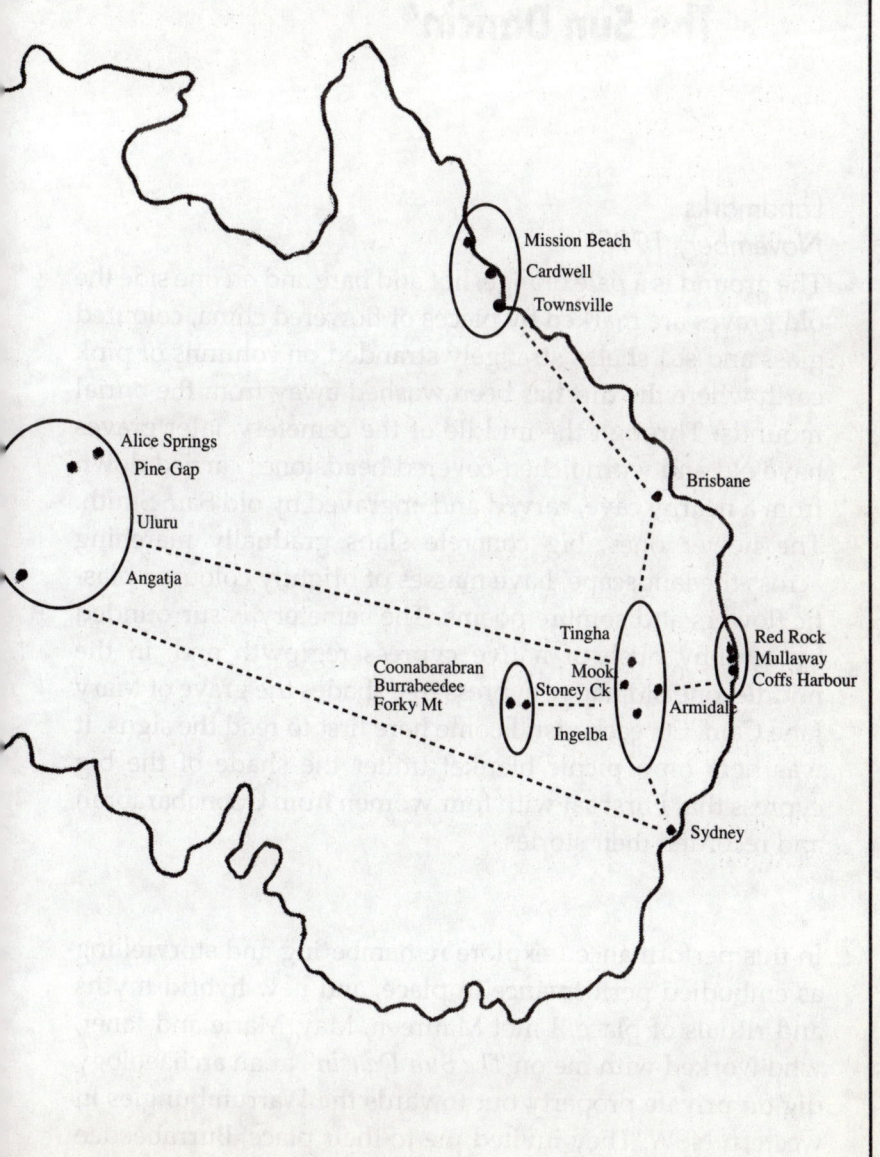

Performance IV

The Sun Dancin'

Landmarks
November, 1993
The ground is a pale orange, hot and bare and on one side the old graves are marked by pieces of flowered china, coloured glass and sea shells, strangely stranded on columns of pink earth where the dirt has been washed away from the burial mounds. Through the middle of the cemetery, later graves have old and worn lichen-covered headstones carried down from a nearby cave, carved and engraved by old Sam Smith. The newer ones, big concrete slabs gradually marching across the landscape, have masses of brightly coloured plastic flowers and sombre poems. The cemetery is surrounded by scrubby bush of native cypress regrowth and, in the middle, one old native cypress tree shades the grave of Mary Jane Cain. On each visit I come here first to read the signs. It was here on a picnic blanket under the shade of the big cypress that I first sat with four women from Coonabarabran and recorded their stories.

In this performance I explore remembering and storytelling as embodied performance in place, and new hybrid myths and rituals of place. I met Maureen, May, Marie and Janet, who worked with me on *The Sun Dancin'*, at an archaeology dig on private property out towards the Warrumbungles in western NSW. They invited me to their place, Burrabeedee Mission,[1] the next day, where we negotiated the telling of their stories. They each defined themselves in some way in

Mary Jane Cain and grand-daughters, Coonabarabran, early 1900s
(Kathy Hinton's collection)

---
1 Although the people referred to Burrabeedee as a 'mission', it was freehold land granted to Mary Jane Cain and not subject to the rules and regulations which applied to Government reserves and missions.

relation to the old Queen, Mary Jane Cain, who had started the Mission in the middle of the last century. I take up the idea of the Queen, following on from Emily, to look at what a Queen might mean for women in a landscape that has been, at least publicly, and from a white point of view, principally defined by men. I also explore the idea of constructing new rituals of place, taking up the story of the sun dancing on Forky Mountain, a new ritual that arose out of the folds of the landscape at Burrabeedee.

~

> ... the memories of women. Patiently transmitted from mouth to ear, body to body, hand to hand. In the process of storytelling, speaking and listening refer to realities that do not involve just the imagination. The speech is seen, heard, smelled, tasted and touched (Trinh 1989: 121).

Margaret: Tell me how to put it in the beginning.
Marie: You put it how you wanta.
Maureen: Say 'Long, long ago in the dreamtime,' eh (laughing).
Marie: We don't wanta tell you how to do it, Margaret. We want you to to do it and bring it back to us.

Thursday, 4th November

I carry winged words on paper along the familiar lines of my journeyings from the New England Tablelands to the Warrumbungles mountain country. These winged words will fly with their story out into the world. It has been a long time, nearly two years of waiting for the contracts to arrive. I retrace the familiar track along the flat tabletop of the New England ranges until I slide down off the edge of the Moonbis into the fertile plain of the valley below. Through the traffic of Tamworth and then turn left. Westwards, and the world opens up; mind expands outwards through the third eye, the pineal gland, centre of meditation. It is a grey brooding day, the outline of the horizon inky blue and rain hanging heavily in the air. An expectant silence reminds me of the words that struck me from Oxley's diaries about his journey into the Warrumbungles in 1820:

> August 9. The fires of the natives were seen at no great distance from us; and they seem to attend upon our motions pretty closely.
>
> August 11. The natives continue in our vicinity unheeded, and unheeding: even the noise of their mogo upon the trees is a relief from the otherwise utter loneliness of feeling we cannot help experiencing in these desolate wilds (Oxley 1820: 263–4).

These diary entries are the only record of people in the area prior to a white settlement which almost erased the Aboriginal story. Apart from this there is the presence of artefacts, the archaeological record. I read the archaeological accounts as a net which is 'a collection of holes tied together with string' (Diprose and Ferrell 1991: viii), and attend to the absences as well as the presences. The archaeological story plots isolated findings of material objects over a ten, twenty or sixty thousand year period but cannot tell the story of the hand that forged them, the life that was lived around them, and of course tells nothing of the cultural genocide which must have taken place. There is a deep and profound silence about what happened between 1820, when Oxley forebodingly describes the silent scene 'in these desolate wilds,' and the memoirs of Mary Jane Cain, shepherdess and Aboriginal Queen, of a pastoral era. As Maureen puts it, *it is the pain of not knowing*.

My meditations induce a deep reverie out of which I am startled by a sense of the presence of people-spirits: Janet and May, Maureen and Marie, Charlotte, Ethel, Madge, Laurel, Gladys, and others, all the people who told all the stories.

> The story depends on every one of us to come into being. It needs us all, needs our remembering, understanding, and creating what we have heard together to keep on coming into being. The story of a people. Of us, peoples (Trinh 1989: 119).

The contracts have arrived from Aboriginal Studies Press who are to publish our story and I take them to the four women, Marie Dundas, Janet Robinson, May Mead and Maureen Sulter to sign. And I am aware, at this moment, of the presence of Mary Jane Cain who made the stories possible. The four women who worked on the story all related their authority to Mary Jane Cain who had crossed the boundary between private and public. Only the very oldest women remembered her—Ethel, and Violet, her grand-daughters, and Emily, her great grand-daughter, talk-storied her into being. Others remembered stories they had been told.

> My great-grandmama told my grandmama the part she lived through that my grandmama didn't live through and my grandmama told my mama what they both lived through and my mama told me what they all lived through and we were supposed to pass it down like that from generation to generation so we'd never forget. Even though they'd burned everything to play like it didn't ever happen (Jones cited by Trinh 1989: 122).

Most of all, everyone knew the story of Mary Jane Cain. She grew up as shepherdess on properties in the surrounding district and, as a young newly married woman, came to live at Honeysuckle Point on the edge of town. There she kept her goats which wandered daily as far as Forky Mountain and every day Mary Jane Cain walked fifteen kilometres each way to shepherd them back into town. Eventually Mary Jane decided that she would build a shack out at Forky Mountain and live there to save the daily walk. She then *wrote to the Queen* and asked that the land be granted to her, which it was, on the condition that she set up a sanctuary for her people.

Like the five matriarchs of Ingelba and Maryanne Sullivan, who were Mary Jane's contemporaries, and Nganyinytja, a modern day leader, Mary Jane Cain is an in-between woman straddling two eras, the time before and the time after white settlement. These were the women who made it *good for ourselves to go forward*. And like the five matriarchs, her motherhood and her status as belonging to the place, were the two critical elements in her power. The people say she had many children and these children had many children; she brought up others' children (a necessity in such times of great distress); and when she *made the Mission*, many people came from far and wide to settle on this land at Forky Mountain. Laurel's parents were among the first to come there:

Laurel: When my mother and father had a little bit of an argue, Dad would say to Mum 'Oh well, you only come from Coonamble with your goats.' I'd say tell me a bit about how you come to be there [at Burrabeedee]. He said 'Walk through the friggin' Pilliga Scrub' you know, like that, walk through the Pilliga Scrub carryin' their swags and all the billy goats draggin' behind (all laughing). See, well they came from Coonamble with the goats and old Granny Marny and old Aunty Ada too, 'cause she got old 'Arry Fuller. Apparently they must have all come together, the Coles and the Marnys, and went to the Mission to get permission off Granny Cain, 'cause she let everyone in no matter who it was. White fellers, dark women with white men—they were quite welcome 'cause she wanted to make a mission more or less as a refuge for them to live with the families, rear their families.

When I think of Mary Jane Cain now it is the image of two photographs of her that I have in mind. Both are distinctive Victorian studio photographs with a Victorian interior as background and figures dressed in the formal Victorian fashion of the time. In one she is surrounded by three of her grand-daughters—touching, loving, close, the warm circle of generations; in the other she is with two grandsons who are in World War 1 uniform—separate, serious, upright, speaking of authority. In these two images I see the two faces of Mary Jane Cain.

It is her (hand)writing, however, that connects me to her flesh and blood. I sat in the Mitchell Library[2] and read with

---

2 The State Library of New South Wales.

sympathetic disappointment her *Recollections of Coonabarabran*, written in about 1920 about the white history of the early period of Coonabarabran, a topic with which Mary Jane was intimately acquainted from her experiences of working on their properties. It was on these properties that she learned to write. The people were disappointed that she had not recorded any of their history but then I ask, what was it possible to write of Aboriginal history in the 1920s? What connections could be made between words used to construct a history of white settlement and the fleeting oral stories on its margins? It is amazing enough that she wrote at all, and that her words survived.

She did include some small autobiographical fragments in which she describes her mother as a *full-blooded Aboriginal who was brought* up in the Mudgee area by the Cox family. Her father was an Irishman, Eugene Griffin, and her parents moved between Mudgee and the Coonabarabran district when they became travelling hawkers:

They had a small cart what would be something like a village cart drawn by two good horses. They used to get these goods from a firm in Maitland. They travelled from place to place from here to Maitland till at last they were overtaken on Walders Range by some Bush Rangers who's names were Long Tom, the Jew Boy and Oppossum Jack. Mother and father not having much cash the bushrangers were quite satisfied to get the goods or whatever goods there was. Nothing serious occurred. After this my father came back and settled down at Bomera as mentioned in another part.

Jinnie and Eugene Griffin spent a considerable part of their lives working on stations and properties around Coonabarabran. Mary Jane Cain says that on one property her father worked as a dairyman, on a farm which grew wheat and tobacco, and her mother was shepherding the sheep. They had been working on a big property, Toorawandi, for some years when Mary Jane was born in 1844.

At the end of this manuscript about white settlement, there was a separate page, unconnected, with a list of Aboriginal words and their meanings, signed *Mary Jane Cain, Queen*. It was in this domain, the domain of Aboriginal language, that Mary Jane assumed her queenliness. The bridge she had used to connect the world of Aboriginal stories with the permitted ones of white settlement was place names, the only Aboriginal language words which had been retained, appropriated to name places significant to white settlement.

It was not this formal piece of writing, however, which led me to whoop with delight and tugged at my heart strings, but a letter I discovered in the bowels of the State Archives. I assumed this unsophisticated and haltingly written letter, dated 1893, was what the people referred to as Mary Jane's *letter to the Queen*. It was actually addressed to the Governor, the Queen's representative in the new colony. The relation between hand and pen is halting, the spelling so unusual at times as to defy interpretation and the story heart-rendingly simple. When I showed a copy of this letter to Violet on my way through Gunnedah one day, she tilted her head to the side and smiled *Granny loved writing. I can remember her sitting at a little desk writing with a feather*. In this letter, Mary Jane makes a very early claim for land rights, on the basis of having nine children to support and being *the only native from around these parts*. She asks for the Title Deeds to the land at Forky Mountain so she can build fences to keep off neighbouring stock.

The seven hundred acres which eventually became Burrabeedee were granted during the 1890s in three titles, and *the power of writing* was critical in gaining title to the land. Queen Victoria was the ultimate authority in the new colony during most of Mary Jane Cain's life. It was Mary Jane's ability and determination to write, in the form of the letter, that spun the threads of The Story from Mary Jane Cain, Aboriginal shepherdess, 'mother of nine' and 'the only native around these parts', to the public authority of the Queen. Over the years, between her earliest representations and the establishment of Burrabeedee, Mary Jane Cain's authority grew until eventually she and others considered herself the *Queen of*

Burrabeedee. An episode, reported with paternalistic good humour in *The Bush Brother* of 1909, illustrates the seriousness with which Mary Jane and the people regarded her authority. Queen Victoria had just died and the anthem had become God Save the King, but the people asked that it be sung again for the Queen:

> A Scotchman would have laughed had he been lucky enough to be present at a concert and dance we had at Forked Mountain in January. We finished up dog tired, with singing 'God Save the King' and when the Queen of the camp begged us not to go without singing 'God Save the Queen' in her honour,—why we did (*The Bush Brother* 1909: 154).

According to Sidonie Smith:

> Sometimes, however, phallogocentric discourse has permitted women powerful life scripts, such as those of the queen and the female religious. As a result, the autobiographer may commit herself to a certain kind of matrilineal contract, tracing her story through a series of powerful foremothers (Smith 1987: 55).

Mary Jane Cain has actually created such a story, borrowing from the practices she observed around her, the discourse of the queen. She is the beginning of a new story through which her daughters and her daughters' daughters trace their author/ity.

In this place of spirits I notice that the landscape has changed, the patchwork of fertile crops and pastoral country giving way to a wildness where native cypress crowd the roadside with wattle and ti-tree scrub in the pink sandy soil of Burrabeedee. Soon I catch a glimpse of the double top of Forky Mountain, signifier of place, and I look for the gate to Burrabeedee. I always visit the cemetery on my way to Coonabarabran, to read the signs in the landscape.

There are new car tracks in the wet dirt leading from the open gate to the cemetery, the people always come here too. I stop at the cemetery and sit still for a minute, a sigh of pleasure escapes, the pleasure of return and I recall Madge:

Madge: Yeah, well see people come back now after years, even people that's been living in Sydney, like poor old Lisa Golden, and them, they all come back 'ere. And when they all come back, they come back to bury the old people and all the young people and we all meet one another all over again. Even my people—when my brother's son died here so many months back—all his sons and that came from down in Melbourne and them places, kids you've never seen for years, never ever seen in your life, till you seen 'em, then—'this is your aunty'. You meet all these people. So that bit's the cemetery, they come home.

Two years since I've been here and there's much that I can read and remember from the graves. It's been raining heavily and most of the curious pillars of pink earth which support stranded decorations of coloured glass and pottery fragments on the very old graves have washed away. There are a few left, marvels of lichen and fungus lace where the ornaments have blended with the soil structure into a sculpture that is neither animate nor inanimate. I wonder whether the people have forgotten the graves but then I remember being here after heavy rain when Maureen and Marie immediately set to, digging, clearing, tidying the graves, talk-storying as they go, about the old people. The newer style of graves, huge concrete slabs, seem to be marching ever onwards towards the old area and I am a bit sad to see sea shells and glass fragments embedded in concrete. But Burrabeedee is about change, it's a story about the negotiation and re-negotiation of space. The old Queen's grave in the middle of the cemetery under a cypress tree is still unmarked, and its miniature white picket fence is lying down beside it. I sat here with the four women when they began to tell me their stories:

Maureen: At least one thing, I say we're direct descendants of the first people, me and you, telling our story about what happened—to them three women there (nods towards the photograph), the three women after Mary Jane Cain, from the three of them that's where they [all the people] come from.

Looking further round the cemetery, I notice that Gladys, sad and withdrawn when I last saw her, now lies next to Joe her husband, their mounds of earth covered with masses of bright plastic flowers. Near to them is the tiny grave of Kylie's stillborn baby, Jacob; but I have heard she has had a healthy baby since. And I pass Maureen's sister Nancy where Maureen and I sat two years ago in the hot midday sun, tears streaming, as she said *how can I remember, how can I forget*, when her body was so cut up that they wouldn't let us see her to say goodbye. Her youngest sister was killed in a car accident and Maureen's first consolation was Nancy's actual words in our book, her second that she had decided to write her sister's story.

> And never forget that writing is as close as we get to keeping a hold on the thousand and one things—childhood, certainties, cities, doubts, dreams, instants, phrases, parents, loves—that go on slipping, like sand, through our fingers (Rushdie 1985: x).

The cemetery enables a reading/writing practice embedded in place.

It is fifteen kilometres from the cemetery to the town of Coonabarabran and as I drive I imagine the people walking in and out of town, especially the kids going to the pictures, scaring each other on the way home with ghosty stories. The Red Hill ghost, a woman who carried her head under her arm, Mother Tongue Bung with a long tongue which curled right down Forky Mountain to gather up naughty children. As I get closer I am aware of the grid of town streets in my head, marked by the location of the houses of the four main women. Two on the north side and two on the south, dotted amongst all the other houses in a gesture that preserves the ethos of assimilation. When the people moved to town in the 1950s, it was so they could learn to be more like whites and the women had learned to construct their desires in this direction.

Burrabeedee, although independent, had a Government, and even a missionary presence, at the invitation of Mary

Jane Cain, because she wanted the children to be educated. It had become the special challenge of the authorities to encourage people to move off reserves and missions and into towns. Reports from the Aborigines Protection Board of the 1950s commented glowingly on the number of people who had moved from Burrabeedee into town, and in 1956 the school was closed. Charlotte told me that she left Burrabeedee so her children could be educated in town. It was the dominant storyline about moving off the Missions and the need for assimilation that the people took on for themselves. There seems to have been extreme pressure for the women to conform in dress, housekeeping, childcare and, dotted amongst the houses of white women, it was assumed that they would learn to erase the traces of place memory, just as traces of occupation—houses and shacks, wells and water lines—were erased from Burrabeedee:

Margaret: What happened about the houses being sold?
Charlotte: Oh well, we couldn't do nothin' about it, it was just their business, the Manager's, after we left there, the fellers come in and bought the homes.
Margaret: But what about—how could they sell all the little bark huts and humpies and everything?
Charlotte: I don't think they did that, they just pulled them all down, just mainly the other ones, half a dozen I think it was.
Margaret: 'Cause they would have sold the teacher's house?
Charlotte: And the Manager's home—that's six.
Margaret: But how could they pull down the others?
Charlotte: Well, pulled 'em off the rocks, put them on the back of trucks, you can see 'em on the properties.

And yet everything is not erased. Each time I visit Burrabeedee with the women there are more stories of their lives there. Maureen performed her story on several occasions in front of the gateposts to the house she grew up in at Burrabeedee.

Each time the performance is slightly different depending on the season, the company and the context of Maureen's life at the time. The effect is one of layering, layers of story, beginning with the most basic description of their garden and house:

This was my gate in 1953 when I left here. Nanna used to have her gardens on either side. Yes, see there, she had a lovely garden on that side, and on this side too they planted a little cedar tree so we could put our swing on it. We used to have a swing on that limb there see.

On another occasion when Nancy came out with us, the gate was incorporated into remembered stories of their childhood:

Nancy: When we'd do something wrong we'd come out the front gate here, come out in our nighties and go underneath the hedge. 'Where'd these two go, oh they've gone, got their bag and gone, runnin' away' (laughing). Daddy said 'Don't worry about 'em then, they won't be long, as soon as it starts gettin' cold, they'll be back.'
Maureen: As soon as it gets dark.
Nancy: Then Daddy'd come out and get us eh, 'Come on you two it's past nine o'clock, in to bed' (much laughing). Then we'd pick our little swag up and go back in and get into bed.
Maureen: (laughing) We would make out we were runnin' away from home. When you're little you have funny but happy times.
Maureen: That's where the front garden was, on either side here. And all down outside there was a big vegetable garden then.
Nancy: We had chooks right down the back.
Maureen: They were good times out here.

And in another performance Maureen staged her front gate as part of a mock landrights ceremony, revealing the deeply held significance of the site for her:

Do you want me to step it out and say 'Don't come on this land' eh? (laughs) I don't want anyone to touch this, I mean it, no matter what, because I'll get on there with m' little Aboriginal flag and I'll say I'm claimin' this bit of land. We want this piece of land so that nobody touches it. I'll take this here.

She contrasted herself in this with her friend Rosie Waten with whom she shares a strong commitment to their memories of Burrabeedee. But for Maureen, unlike Rosie, those memories also involve her politicisation:

Rosie'll never forget what happened out here at the Mission either. She never got involved in Aboriginal affairs and things like that. Not like us. We were involved—Marie and m'brother Auby was involved. My father couldn't understand why I was doing for people that didn't appreciate anything I was doing for 'em, and all I could say to him was that's how I'm made, it must have been God's way for 'im to say well you do this.

Maureen is still highly political, has completed teacher qualifications and is currently completing a degree in counselling. She bases the authority to tell her story on her relationship to the land at Burrabeedee. This time I have left her behind in Armidale doing her study.

~

The people have been living in town for forty years now. It is my habit to visit the four women first, to let them know that I have arrived and to arrange a meeting time. Marie is first, at 24 North Street, where husband Kenny and son Phillip answer the door with grandchildren in tow. Marie is, as usual, down the street, always at bingo, or the TAB. She never learned to conform, is a self-confessed gambler; wilful, and insistently oral, she is intent on disrupting authority. It was Marie who coined the metaphor of *the pencil and the mouth* to talk about the relationship between speaking and writing, orality and literacy:

I'm not an educated person got a pencil, understand. I could stand up in a crowd and tell them what I think but if someone says 'Marie you write that down,' I can't. Jill would always sit back and she'd say 'I'll do the pencilling and you do the mouth 'cause your mouth can win, you're like if you're in a cave.' At the Lands Council

Meetings—I went to one in Dubbo with Kenny and they didn't want to give Coona money, so there was eleven or sixteen gettin' funded and not Coona. So anyway, I stood up and said, 'Well, why not Coona, they've all fiddled the till, not only one person, they've all done it and yous can give 'em another go.' I turned round and I said to Kenny 'Now don't you use me either, to tell 'em,' and 'e said, 'You keep goin', bub, you're doin' a fine job, I'll write it down and then I'll flog the bastards,' and 'e did.

She told abundant stories about growing up with twelve brothers and sisters, up behind the Showground, *we were the Showground kids*:

We built the house of whatever we could find (laughing). If we had a piece of good tin, we'd think we was the best kids in the scrub. Cardboard—we built it out of tin first—then they'd go chop the poles. You know how they make it the shape, then the tin on the roof, then we'd go down to Billy Neilson's at the goods shed and get every cardboard box 'e 'ad, then buy a packet of tacks and Daddy 'ad the hammer. That was for the walls, that's our walls, cardboard walls—but real thick cardboard, it was good cardboard. Only time if it rained too much we had to throw the walls away and go down the goods shed and get more walls (much laughing).

Marie is both interested and sceptical of my talk about questions of authority and the Queen and one day asked her mother *who was the Queen now*. While on the one hand she suggests that her mother, as the oldest woman from Burrabeedee, should be Queen, she also teases her with the idea of Charlotte, a slightly younger contemporary, being Queen, and effectively disrupts the whole idea of authority:

Ethel: Queen Robinson, she was the Queen [in Ethel's time].
Marie: But she's not the Queen now.
Ethel: Granny Cain was the Queen, you said.
Marie: Charlotte's the Queen now.
Ethel: Charlotte's not the Queen.
Marie: Well they call her the Queen, she's the queen bee.
Ethel: She's not the Queen.

Marie: Who do you want to be the Queen?
Ethel: She shouldn't be the Queen.
Marie: (bursts out laughing) Do you want to be the Queen?
Ethel: Course I don't want to be the Queen, 'cause I'm not the Queen.
Marie: You don't want Charlotte to be the Queen (laughing).
Ethel: Charlotte's not the Queen.
Marie: They say Queen Charlotte.
Ethel: Queen Charlotte (disparagingly).
Marie: Everybody in the town bows down to Charlotte, everybody, all the white people and they all bow down to Charlotte.

Marie, determinedly oral, defines herself as on the margins, and appears to have no need for authority. She is aware of her connection to Burrabeedee, and Mary Jane Cain, and interested about the idea of Queen in relation to her mother, but throughout my work in Coonabarabran I could rely on Marie to disrupt the seriousness of any idea.

I find May in her new house, small, neat and clipped on the outside, but awash with toys inside while she looks after her young grandson. She moved last year after her husband died and we spend a while catching up on his long drawn out illness. The first year was the hardest she says, but now she feels well and strong. She has given up her charity work at Vinnie's[3] but works on the hospital committee. May always seems to be busy and involved in the life of the town but her happiest memories are of roaming the bush in her earlier days:

That was a part of my life that I really liked, when we were out walkin' around the mountains, and catchin' kangaroos. There was goats and if we needed meat we'd get a young goat. We didn't kill the old ones, we'd get a young one and kill it and that was good. They didn't have much, the men, because sleeper cutters only get paid once a month. You had flour, sugar, tea then you'd catch your

---

3 St. Vincent de Paul is a charitable organisation staffed by volunteers which sells secondhand clothes and provides food and furniture relief to people in need.

own rabbits or goats or whatever for the meat part of it, and buy onions and potatoes which weren't as dear as they are today. They'd set traps. Clara used to set traps and catch rabbits for food. We had one lady out there that used to be with the sleeper cutters, she used to make kangaroo rissoles.

She is excited by a new discovery that she is related to Mary Jane Cain, 'I can claim a bit of Burrabeedee too now,' she says. An interesting discovery because May had always defined herself by her exclusion from Burrabeedee, always an outsider in Coonabarabran. Every story has its exclusions and inclusions, voices and silences, and for May the exclusions of the Burrabeedee story were harsh. May came from Baradine, just thirty kilometres away, but her family were never accepted as 'Burrabeedee' although they knew all the Burrabeedee people and visited there often. Excluded from the authorship that Mary Jane's story conferred, May claimed her authority on the basis of her brothers' and uncles' participation in the wars and on her own participation in the workforce under the Manpower Scheme. This new discovery, based on a complicated genealogy, enables her to rewrite her (hi)story. Several of the women, including May, take a keen interest in genealogical research and there appears to be as much dispute about written genealogies as Walter Ong notes about oral ones:

> In recent years among the Tiv people of Nigeria the genealogies actually used orally in settling court disputes have been found to diverge considerably from the genealogies carefully recorded in writing by the British forty years earlier ... What had happened was that the later genealogies had been adjusted to the changed social relations among the Tiv: they were the same in that they functioned in the same way to regulate the real world (Ong 1982: 48).

For the women written genealogies are subject to similar relations of power and they are also always subject to the stories of memory. The women are less interested in distant genealogies but have stored in their memory hundreds of rhizomatous relationships which spread out over up to five generations. In *Ingelba*, Patsy represented these as concentric circles originating in the five matriarchs. In Coonabarabran, they also relate very much to Mary Jane Cain as origin and also as authority.

Janet comes out when I knock, big, round and dark, and flops on a deckchair on the front verandah to talk. I notice immediately, with a shock, the space where her right breast should be and Janet places her hand there and tells me the story. Two operations in Coona and six weeks of radiation therapy in Sydney. All the women in their yellow gowns, not knowing whether to look or not, and they were all in there for the same thing after all, she says. They were a sorrowful lot, my mob laughed all the time, but the others were quiet, a sorrowful lot. I was the only black one there, I didn't see any other Aboriginals. When she left the doctor gave her a box of chocolates and a kiss and said that's what I do for my special patients. I'm glad to be home, she says.

I think about what gives these women the strength to endure and I wonder if it is The Story. Janet has always felt inadequate about her lack of 'tribal' knowledge but defines herself and what she knows as the story of her great grandmother:

I can remember when I was a girl we weren't allowed inside while the grownups were talkin'. We had to go outside and play cubbyhouse. Or we'd have our meals first and we'd have to leave the table and let the grownups sit down. So we don't know anything about blacks, really. Cause my Aunt Joffie, she wouldn't eat unless she had a tablecoth on the table.

When we went down to Shepparton we were sittin' down with this old tribal feller tellin' us stories about how they used to take 'im out trackin' 'You'd never starve,' he said. I said 'We will, I wouldn't eat a snake or a goanna or anything. I'd run a mile.' We weren't taught that. The only thing I read about round Coonabarabran was The Red Chief *book and that's the only thing I ever read about Aborigine. We couldn't live that life, we'd die.*

I don't know any blacky stories. All I know is the Cain—you know, how we was reared up and the mission was given to my great grandmother.

Janet's authority is very much based on her relationship to Mary Jane Cain through her mother and her grandmother.

'They're going to publish the book,' I say and Janet finishes my sentence 'after all these years'. 'What about the name?' she asks, recalling my last letter telling the women that the final hold-up was a disagreement over the title. 'That's all sorted,' I say, 'I told them that the sun dancin' story was the main story that came up, over and over again, and that we had thought of it as *the sun dancin'* for so long we couldn't change it.' 'Yes,' Janet nodded in agreement, 'and it's only just about our story that we know.' All the women told me the story of the sun dancing:

At a certain time of the year, around Easter time, Daddy and Granny and them—they used to always go up on the mountain at Burrabeedee and watch the sun dancin'. Then they'd decide to take us but we had to be good to go, which we was then in them days. I didn't know what it was, then Daddy said 'Just watch it now and it'll dance for you,' and it used to dance and the different colours, pretty. A lot of people don't believe in it but it does, it does do it.

It was just ordinary, then it was yellow as it was risin', then all of a sudden the colours started comin'. I don't know whether it was a reflection or something, then all of a sudden it'd start doin' this (shimmering), it dances. Daddy used to say 'It's dancin' now, look!'

I puzzle about how a story like this comes into being, why it became a yearly ritual, and its possible links with a Christian tradition. Christianity was brought to Burrabeedee by two old missionary ladies who were much loved and the people built a church so they could hold their own services out there. Christmas and Easter were important occasions of celebration. Christmas was celebrated with typical, grand scale feasting but Easter is only mentioned in the context of the sun dancin'.

The ritual was apparently carried out in other parts of Australia and is mentioned in a poem by Mary Gilmore (1931) in which part of her own ritual of place on the Gundary Plain was to *see the sun dance up on Easter morn*. The footnote to this line suggests that when they were very little children their parents used to wake them up to see the sun dance on Easter morning and that Russia is the last country in which the custom is retained, though no longer officially (Gilmore 1931: 15). It is likely then that the custom pre-dates Christianity and may have links to the pre-Christian festival of Ishtar, when the beginning of the year was celebrated at the time of the strongest moon. As an Australian landscape ritual it presents new possibilities for envisioning the landscape and I imagine a pilgrimage to Forky Mountain to watch the sun dance on Easter morning. Again, I am in the space in-between and I like the play of hybrid richness of these layers of story in place:

> Language is the site of return and the insisting call from afar back home, but language is also the site of change, an ever shifting ground. To borrow a wise man's words, it is constituted as an 'infinitely interfertile family of species, spreading or mysteriously declining over time, shamelessly and endlessly hybridising, changing its own rules as it goes' (Trinh 1993: 4).

I can't get all the women together to decide on a time and place to meet and there is no designated Aboriginal meeting place in Coonabarabran. May will be away all day tomorrow so I take a punt and suggest that we meet in Cardian's Coffee Lounge the next day. I then retreat to the little cottage in the bush where I am staying.

At night I am tired and sleep in my cubbyhouse bed held close by silvery trunks of eucalyptus rossii and lulled by the beat of rain on a tin roof. I dream that I have climbed to the top of a mountain. It is a long way and a hard climb and I balance on the top looking down the other side. I am so tired I feel in danger of rolling all the way back down again so I sleep on top of the mountain until I am ready to go back again.

Friday, 5th November

I have a day off because I decide not to visit any more people until we sign the contracts. Inside The Cottage with pot belly puffing to fend off cold and rain, I read *The Sun Dancin'* manuscript and think about my original question:

How can I have a sense of belonging in the Australian landscape?

Who am I in relation to these women and their stories? Is it four women or five? I puzzle over these questions for some time and one day when I am walking along a bush road in Coonabarabran an image of my presence comes to me. My consciousness is moving in and out from thinking about the book to the texture of the gravel road under foot, the clouds flickering across the early morning sky and the cool lightness of air on my face when I notice another presence. This other presence, elongated and slightly distorted in the characteristic way of a Drysdale bush woman, walks with me. The sun is rising in the Eastern sky behind me and has cast my shadow on the ground in front in what seems a powerful image of my place in this whole process.

~

I was thinking about this question of belonging when I visited my Aunty Betty last Christmas. While I was with her she opened a Christmas card which had a fine ink drawing of a lake surrounded by fir trees and hills on the front. She looked lovingly at this drawing, and identified it as a lake in Scotland. Surprised, I asked her how she could recognise it. She told me that my grandma had talked to them all through their childhood about Scotland, describing in fine detail the Scottish landscape. She always called Scotland *home* and visited every second year. She said Grandma hated the Australian landscape, seeing only not-Scotland, not-home.

I hadn't realised until then how much a migrant I was. In listening to the *Sun Dancin'* stories, I was inserting myself bit by bit into this landscape:

I am the listener, present in all the stories. I am there in the negative image, a shadow on the landscape of what is happening. My eye positions the camera but it cannot take a photograph of me. The only way I can observe myself being there is in my journal writing, my reflections on what is happening at the time. Reflections, shadows, moments of being captured from the inevitable flux of daily living. 'Shadows define the real ... I see them as making some sort of sense of the light' (Dillard 1974: 62). My presence changes like the shadow, long and ephemeral in the morning sun, growing solid and dark towards noon and flickering in the light and shade. Now you see it now you don't.

I see myself only as a shadow in the landscape of the writing of *The Sun Dancin'*. How can I understand the absence of the body, my body? Many feminist theorists have engaged in re-writing the body, most particularly Liz Grosz whose project of 'having a fling with the philosophers from the point of view of the body' has covered many years of her work. Her work, although very illuminating, refers firstly to philosophy and provides no immediate answers for me now. Another theorist, Drew Leder, also asks the question 'Why, if human experience is rooted in the bodily, is the body so often absent from experience?' (1990: 69) Then, unlike Grosz, he suggests ways, other than illness, of imaging a bodily presence:

> Inside and outside, self and Other, are relativized, porous, each time one takes a breath. The air is constantly transgressing boundaries, sustaining life through interconnection (Leder 1990: 171).

There has always been a sense that the talking of the stories has at least in part constituted the stories themselves. Patsy said to me during the recording of the stories from Ingelba, 'I have never spoken about this to anyone before.' Together we spoke the stories into existence, they depended on us speaking, on the breath. And on the pages there is something about representing the words as spoken that makes them present on the page more so than if they were just a secondhand account of their words. They somehow contain the rhythms of the breath, the traces of the body of each storyteller. Leder talks about this breath, about breathing as the site where boundaries between inside and out are transgressed.

Over the course of my work in Coonabarabran, recording and transcribing stories, negotiating them in print, I found that there was no longer a division between 'me' and 'them'; I stopped using the words *Aboriginal people*, and simply wrote *The people*, and their stories became The Story. I say this with a full recognition of the hegemonic potential of this awareness, but needing some way to express the movement that listening to and telling these stories made when they got inside my body. Many of the stories were told in the landscape of Burrabeedee at Forky Mountain and the telling provided the context for more experiences and more stories of which I became an intrinsic part. I found myself with a growing sense of belonging in that landscape, I felt comfortable going to Burrabeedee alone, felt that I too could talk to the spirits there and that I was drawn to and enfolded in that place. Through listening, telling and writing the stories of Forky Mountain, I belonged to Forky Mountain in this sharing of the breath.

When it came to how this question was perceived from the outside however, it was a different matter. The problem was articulated as one of crossing boundaries, the danger of the margins. It is neither a women's studies text nor an Aboriginal text, we don't know how to market it, they said.

> Across the lines who would dare to go, over the bridge and under the tracks, separate white from blacks, two sides, running for our lives (Tracey Chapman song).

The general opinion seemed to be that I should erase my presence in the text. That it was better to pretend that the stories stood alone, that the circumstances of their production weren't as they were, but that they somehow got produced by themselves, without bodies and all the politics of their production. Or that only Aboriginal people should write the stories themselves. And yet, I argued, both for myself and for potential readers, if there is no chance of having an embodied presence in the landscape of these stories, then how are we to develop a relation of empathy (see Probyn 1993b—Performance 11) to the tellers? How can we understand the magnitude of dispossession and the possibility of new stories? Or are we to be forever aliens in this landscape of Australia? In Marie's words, *we've gotta make it good for ourselves to go forward.*

> I sometimes wish I could be
> one of The People
> then I could tell these stories differently.
> Instead of marking the separation carefully
> with text and punctuation,
> always showing, knowing, that I am Other,
> I would abandon careful separation
> and be there fully,
> tell the stories I love so much
> as if they were my own,
> take on their resonances,
> the turn of phrase I know so well.
>
> When I sit at night with Kathy
> talking at the table
> and silent in the morning
> by puffing pot belly
> soaking up warmth of fire
> and her large presence
> I do not feel separation
> I feel belonging,
> *Être-là*, being there.
> I try to work out
> How to insert this bodily presence
> into the text.
> But then the publishers say
> You are crossing boundaries.
> You must decide whether
> it is an Aboriginal Studies text
> or a Women's Studies text.
> We don't know what category
> It doesn't fit
> our publishing guidelines
> We don't know how to market it.

> We want you to tell a different story
> a story that is not
> broken
> fragmented
> place-based
> postmodern.
> We want a story that hangs together well
> with a continuous narrative
> so we all know where we stand.
>
> Well I can't
> 'cause I was there
> and the important thing to me
> is to be more there,
> not less.

At night, in the fairytale bed in the fairytale cottage, I read Oliver Sacks on the importance of story to our sense of self. He describes patients suffering from brain damage who 'must literally make themselves (and their world) up every moment'.

> He would whirl, fluently, from one guess, one hypothesis, one belief, to the next, without any appearance of uncertainty at any point—he never knew who I was, or what and where *he* was, an ex-grocer, with severe Korsakov's, in a neurological institution. He remembered nothing for more than a few seconds. He was continually disoriented. Abysses of amnesia continually opened beneath him, but he would bridge them, nimbly, by fluent confabulations and fictions of all kinds. For him they were not fictions, but how he suddenly saw, or interpreted the world (Sacks 1986: 104).

I am struck by this account of how the loss of part of the body/brain that is central to the remembering of story destroys a person's sense of relationship to place and to others. Story as body, body as story, body of a story—and how critical story is to our sense of self.

> We have, each of us, a life story, an inner narrative—whose continuity, whose sense, *is* our lives. It might be said that each of us constructs and lives, a 'narrative', and that this narrative *is* us, our identities (Sacks 1986: 105).

But then Sacks' account also seems the perfect metaphor for the power of Mary Jane Cain to provide a new story for her people after their total dispossession. I see loss of land as equivalent to loss of the story remembering part of the brain, resulting in a similar loss of self. Mary Jane creates a new landscape, a new story and a new embodiment for the people.

But what of the second dispossession during the 1950s when a new and powerful story of assimilation was taken on? Did the people lose their story again? It seems rather that The Story became submerged, invisible, always immanent in the landscape of Burrabeedee, rather than lost.

Saturday, 6th November

I worry as I head towards Cardian's Coffee Shop that I have chosen the wrong place to meet and no one will turn up. As I wait I realise more strongly than ever the absence of a meeting place. In being scattered throughout the white town there is no place where we can meet to tell stories. When I can afford it I take a cabin at the tourist village and everyone turns up there to talk. The only other meeting place is the bench outside the Courthouse. I also remember the many times I have negotiated times and places with particular people and different people have turned up at a different time and place. I have had to let go of preconceptions and assume that if I am just hangin' loose, something at least will happen.

I decide to walk down the street to get the paper to read while I'm waiting and Marie shouts at me from across the road. She's sitting on the bench with Janet under *the tree of knowledge*, at the Courthouse, waiting to catch sight of my car. 'Where's the car?' Marie calls out. 'I parked it round the side there, I've been waiting at the coffee shop,' I say. They both look disbelievingly and point across the road at a place called the Lunchbox that I didn't even know had coffee. 'No! Cardians,' 'Oh, Cardigans,' Marie says, 'OK,' and then drives off to find May, leaving me with all the grandkids. Cardigans are already a bit amazed by me and the grandkids when

Marie, Janet and May eventually turn up and Marie takes the floor in storytelling mode. Sitting there in the coffee shop she gestures towards the south with her chin and re/members the *crows nest* when *Daddy brought us kids to town*.

They had the tree guards [benches around trees in main street] *eh? Run to see if you could get a tree guard. Sulky and horses pullin' up, and Billy Chatfield used to always get it, old Billy Chatfield, and Mummy took all us kids down and I could still hear Daddy sayin' no you wouldn't get a tree guard 'ere, Billy Chatfield always gets 'em (laughing). And Queenie Robinson, Aunty Queenie used to take it and then we would go down to the Goods Shed and all the Burrabeedee sittin' there, that's where the mission truck used to pull up and Daddy used to say how could you bring yer kids down town Ettie, the Goods Shed's taken with blacks, Billy Chatfield owns the tree guard and Burrabeedee owns the Goods Shed. We'd only come to town us kids once a month and every time all the kids'd come once a month, endowment day and the tree guards, and you know all them cedar trees, they are still 'ere today, they were 'ere years ago, you know right down to the first part of the clock and I tell you what when the black mothers come to town from Burrabeedee with the kids for the endowment and we come from the Showground, the Quintons, the Colliers and the Sutherlands couldn't get a guard or a perch, we'd have to walk back home.*

What Marie presents is the most finely detailed spatial analysis from her black centre of the experience of coming to town once a month on Endowment Day. She is telling me a story about being at Cardian's. The only place blacks could sit in town was on the tree guards in the centre of the street, or on 'a perch' at 'the crow's nest', the old goods shed. Often all the spaces would be taken up and her father would insist that they go home again. In the days of the Mission, the Mission people had to be out of town by six o'clock and the picture show had a metal dividing fence down the middle.

We sit now, not at the crow's nest, but in Cardian's Coffee Shop, with Marie's stories still pouring out, more confident now of her power as storyteller, 'We might be millionaires, like Michael Jackson,' she says.

Her latest story is about the Deb Ball and continues the contested discourse of the Queen. She tells us that old Mrs Leslie was given the place of honour at the ball because she is the oldest but Marie believes Charlotte and Ethel should be Queen because *they was with the blacks all their lives.*

Over the road where we go to the Aboriginal Arts and Crafts Shop to phone a JP to witness our signatures, the four women walk through to the back room where they continue the story of the Deb Ball with books and books of photographs. We flip through pages of brilliantly coloured ball gowns on all shapes, sizes and ages—and I wonder what this display is all about, curious that these women, all grandmothers are involved in such a ceremonial 'coming out'. Is it a substitute for what they were unable to have as girls; is it a coming into public view in terms of their storytelling; or just an opportunity to have a good time? I have never seen Marie in anything but tracksuit bottoms and T-shirts, even when she is at the club for bingo and here she is in a hot pink satin dress with puffed sleeves, pink and black netting petticoats, high heeled, hot pink shoes to match, and hair in a fifties bouffant with little white flowers. 'Where did you get the bosom?,' I joke with her. 'I made 'em,' she says, 'didn't I, eh Janet, out of foam rubber and I sewed 'em in' (much laughing).

Meanwhile Shelley, the Aboriginal JP, can't come so May has gone across the road to get the woman from the Frock Shop. The woman is quite openly surprised by our request to witness our signatures for the contracts for a book. 'How many pages?' she asks. 'About two hundred,' I say, and she looks even more surprised. We all sign in front of her and she says that the Coona shopkeepers will help with the launch.

The women are initially a little hesitant about such a coming out of their story in Coona. The Deb Ball is one thing, telling your story to each other and to me is all right, to become publically visible in general is fine, but not in one's home town. From our earliest conversations however, one of the main motives for getting a book published was to make their story visible to the local non-Aboriginal population from whom they had experienced invisibility and racism all their lives. But now it is about to happen they talk about the direct effect on their material present rather than the earlier projection of their own desires and fantasies onto an imaginary situation.

The contracts now signed, I spend the afternoon visiting Marie's two Queens, Charlotte and Ethel.

Charlotte looks even more diminutive than I remember her. Her shiny brown arms are thinner and a pink diaphanous dress accentuates her frailty. Nevertheless she tells me she was at a party in Dubbo until three this morning. The bright orange chiffon curtains sway in the breeze across doorways reminding me that Charlotte has always been a partygoer. Until recently she would dance in high-heeled party shoes until midnight at the club and then go looking for another party. At eighty she fell and damaged her hip and lost confidence in her feet. She has difficulty getting around now and moves with a walking frame. But she often remembers dances and celebrations of all sorts at Burrabeedee where bush musicians gathered and people came from miles around. Charlotte listens with great pleasure to her old friend Joe's tape about the old dances:

Joe: We'd have a comb, Aunty Queenie used to play the comb, and we'd have a mouth organ, gum leaves and things. Poor old Ned Fuller 'e was the old violin player, and then we had another old feller that was 'ere, way back when I was young, he could play the concertina and then my Uncle Tom Cain, he could make a concertina talk, I can just remember 'im. They had their own music like concertinas and violins and that too but then as they went away they left us with the comb and mouth organ. Old Jack Bates used to play, years ago now, play for the dances, then they'd start dancin' at 8 o'clock in the night and they wouldn't finish till 8 o'clock in the mornin'. We had great times out there.

There were weddings, sports days, bonfire nights and pretend corroborees and Charlotte ran and danced in them all:

Once a year then they'd have that 24th May on account of cracker night and they'd make a big bonfire in the sports ground. They had a big sports day, she was a great runner, the old lady. They used to have a rooster race and she'd catch the rooster all the time. When it come through in our generation it was still the same then, we'd have to be in there catchin' the rooster, all these women, you know, out there catchin' the rooster.

I leave Charlotte with her reminiscences and head up to the old people's home to meet May and Marie and pay a visit to Ethel; if I go alone Ethel doesn't remember me. Since my last visit Ethel has been transferred to the hospital wing and we make our way along disinfected corridors to a room where corpse-like bodies are propped on chairs around the walls. The smell of urine mingles with antiseptic as vacant eyes, open mouths and un-remembering faces stare into space. Ethel is there in a wheelchair whimpering incoherently to attract the nurse's attention but the nurse is busy administering a routine which maintains a semblance of normality. It is afternoon tea time. Ethel's legs lie uselessly crossed in front of her, unable to walk through lack of use, so we wheel her out to the verandah and face the effort of communicating across a seemingly impenetrable gap. Marie remembers to her—'Who was the most beautiful girl on the Mission when Daddy came out to visit on a horse?,' and—'Remember the Golden Gate Bloomers?'

Ethel and Marie laugh as Marie tells the familiar story of how some of the girls wore bloomers made from flourbags, and when they were hanging on the line you could see the Golden Gate brand name across the bum. Soon Ethel's speech and mind joins us and after about half an hour Marie and Ethel are both storytelling loudly and abundantly, mostly at cross-purposes with each other, unconcerned that the other is not listening and each hoping for an audience from either

May or me or passing nurses. I am convinced that it is only the hearing and evoking of her own stories that brings her alive again, and reflect on the double pain of her life there. No one knows her stories. From time to time Ethel grabs my hand and asks me if I have a car and if I will take her away with me. By the time we leave she is talking politics in the present, persuading Marie to go to the Aboriginal meeting and get one of the large new brick houses being built around the corner from North Street. Ethel will then come home and live with her.

After two hours we have to go. I say goodbye sadly and return to the fairytale cottage for an early night because at dawn I will climb Forky Mountain.

At dawn, the mountain is shrouded in misty cloud, still and silent. Droplets of moisture hang from needles of native cypress and as we climb higher, green lace lichen on loose rocks is matched by long trails of old man's beard hanging from gnarled trunks. In this fairytale green and forested world, I easily imagine old Mrs Tongue Bung curling her enormous pointed tongue down her ancient mountain home to swallow up naughty children.

At times it is so steep and footloose that we wind back and forth across the slope and finally emerge from the mist as we reach the top. Through wisps of cloud we can see far away, as far as Gunnedah, a hundred kilometres in the east. We sit at the top of the lower peak and contemplate the distance. The *fork* in the middle is a saddle, a circular enclosed space of silence and spirits. Climbing to the top of the second peak there is a sudden scuffle in the silence and just over the rise a herd of about twenty goats bounds away, perfectly at ease on the rocky mountain. They are probably descendants of Mary Jane Cain's goats that Violet remembers fetching down from Forky Mountain:

We used to have to go up on the mountain every mornin' in the frost to catch our goats down to milk. When we milked the goats we used to have little stools, fetch them down and then put their legs up on the front, we used to milk 'em like that (laughs). Then they'd wander right up in the mountain, we'd let 'em 'cause of all the kurrajongs, and they'd camp up there. We'd have to go up of a morning after 'em and fetch 'em down and milk 'em.

Another time, as I walk with the women along the base of Forky Mountain in shadow of the late afternoon light, I see dotted lights of translucent white irises among native trees and grasses, lights marking the way to a tiny grave. I feel the inscription on a weathered headstone which would have been carved by Old Sam Smith. It must say 'In memory of' but I can only feel the OF and then CHATFIELD, a common family name, and then 1 MONTH, 2 WEEKS and 3 DAYS.

~

The people were born and died at the mountain. Violet says:

I was born in 1917, I'm not a kid am I. I was born there right under the mountain (laughs). A lady called old Maggy Marney, she was an old—they called them midwives in those days. All the people was all buried out there. They was all born under the mountain there.

Emily speaks about midwives who delivered the babies:

When the midwife would come we never even woke up to the fact and we didn't even know our mothers was big and out of shape, we didn't look. See kids was very innocent those days. They'd send us all away, Aunty Violet specially. She'd take us up for a ride or take us down the end of the Mission and play rounders. When we come back, 'Come in 'ere, there's a new baby in the house.' 'Where'd it come from?' 'Oh it come down off the mountain.' See, we never knew. The midwife'd be there and the baby would be born in our own 'ome.

And Laurel describes laying out a body:

The body was laid out on a table. You'd sit a day and a night wasn't it? You know when they used to lay 'em out on a table. Men would be one night and the women'd be next. The kids, they could come in and sit for a while but they'd only have about five minutes and they had to behave themselves. But not all night. I only remember two and I can't think who they were but I know the women and the men used to sit up all night and take shifts.

I know Auntie Queenie done one there, once. Molly was there, Molly and Julia. I don't know who it was, but they had this person laid out there on the table in her lounge room. In her front room, they called it front rooms then. A big old wooden table. They used to make the shrouds for 'em. The women'd do all the sewin', out of calico. And the wreaths, the women and the children used to make them out of wildflowers.

At the foot of the mountain you can see the site of the little shack in which this scene took place. There is an old well-hole overgrown with blackberries.

She put a kitchen extra room on. That was made out of kerosene tins and gal-iron, that was the kitchen and she 'ad two bedrooms and a lounge room but it 'ad a fireplace, open fireplace in it, that was like the front entrance. Only really a two room place and then they put the other room on and the kitchen. It was sort of built up high on blocks. It was only a couple of steps into the kitchen, that was a ground floor. She had an open fire with a big iron across where she used to do all 'er cookin'. Cook 'er own bread, and rabbits. We used to have a lot of rabbits, and now and again we'd get kangaroo tail soup. The other rooms were all boards, the first two—and the one on the other side, they put boards in there but they didn't in the kitchen. And she had old iron beds, with the most beautiful crocheted quilts and patchwork things that they made over the years.

The mountain drew Mary Jane Cain to the place following her goats and the people developed a life there in which they could live their new stories. The image of the mountain permeates many of these stories, the people have re-made themselves out of the folds of the landscape. Kathy even told me that when the women made their Christmas puddings they cooked the pudding in the corner of a flourbag and stood it with the two corners pointing up *just like Forky Mountain*. As Hélène Cixous puts it:

> If we don't invent a language, if we don't find our body's language, it will have too few gestures to accompany our story. We shall tire of the same ones, and leave our desires unexpressed, unrealised. Asleep again, unsatisfied, we shall fall back upon the words of [white]men—who for their part, have `known' for a long time. But not our body (Cixous cited by Grosz 1989: 128).

After the dawn walk up Forky Mountain I make one last visit to see Kathy Hinton. Apart from the four co-authors, Kathy has been the most important person with whom to negotiate the writing of this story. She is self-selected keeper of stories and has been rigorous in keeping me in line. She was expecting me at ten, and at eleven I am in trouble for being late and I explain that I have been climbing Forky Mountain. She nods as if this is expected but it goes unremarked. We sit for a while, Kathy, husband Keith, and Nancy, the only one of nine children still at home, in the tiny lounge room catching up on family news and talking politely until she indicates it is time to move into the kitchen for a cup of tea.

The kitchen is Kathy's working space. She sits at the head of the table and I am at the side; together at the table we do business. Whether we are tape-recording, swapping information from her book or checking through transcripts from *The Sun Dancin'* it is always at the kitchen table and Kathy sets the agenda. She has made the usual plate of elegant triangle sandwiches, a freshly baked cake and the good china is on the table.

Kathy and I have developed a special relationship. I remember when she took me in when I was sick all night out in the Warrumbungles. She warmed me by the kitchen fire and tucked me into bed with flannelette sheets and hot-water bottle. In the morning, before the others got up, we sat together quietly as she coaxed red embers alight in the fire. But this hard won closeness was not without its proper struggle.

Kathy pours the tea and then puts on her reading glasses. I have been waiting for the sign. I get out the manuscript, nervous now, one last but critical permission to achieve. 'I have to check with you Kathy, about publishing your letter,' and hand her the pages:

Dear Marg,
Keith and I have gone through this file [transcript] several times and it's not as I want it to be put in a book for others to see and laugh at. So we have crossed out most of your writing and left what we think you could choose from.
But be careful love. If I sounded as an old black gin I'm very sorry. I know how things change on tape and when you're speaking with others. I don't think I am the right person to direct you in where the homes were. I'd sound much the same and I'd have to cross out the little dunnos, em, e and all that broken English. Shame. Maybe one of the other girls can assist you.
Here's hoping,
Love as always
Kathy.

When I got this letter I felt great pain. Every person I worked with in this community took a different position with regard to the important question of orality, language use and representation. Kathy was at one extreme. When I first asked her to record her stories she was hostile and after she grilled me intensively it emerged that she also was writing a book. We slowly got to know each other on the basis of our mutual writing. She showed me her book, a very privileged reviewing of her life story, tracing the generations back to Mary Jane Cain and down through her grandmother Minnie Mclaren to the story of her own life. She has been working slowly on the book for a long time, writing by hand, adding relevant historical cuttings, photographs and memorabilia to go with her story. She will leave a copy for each of her nine children.

When we had got to know each other more and Kathy was more confident of her own ground, she wanted to tell stories for *The Sun Dancin'*. We recorded some stories alone and some with her mother, Charlotte. Then, as was my practice, I sent the transcripts for her correction and approval. It was then that I received her letter in reply. My heart sank. I was at a particularly low point in the project, finding collaboration with a group of people, and ideally a whole community, a very difficult challenge compared to the safety of my collaboration with Patsy on *Ingelba*. This was especially true as the community remained fragmented and hostile, both within and without, continuing to suffer the alienation imposed during the period of assimilation which resulted in the closing down of the Mission. I understood Kathy's position so well, the struggle to attain acceptance, not only in housework, dress and manners, but in correct writing and grammar, and her perception of local attitudes to language use: 'It's not as if I want it to be put in a book for others to see and laugh at,' and 'If I sounded as an old black gin I'm very sorry.'

Two other women had also talked about blackness and whiteness:

Mrs Leslie: *She 'ad a big white fireplace that you could drag the logs over it, and it had a sort of a hob around it and she used to have it whitewashed every day, white as anything. Then there'd be a fire there next night again.*
Madge: *Go round and dig the whitewash out of the 'ills.*
Mrs Leslie: *They were good too, they used to be white. Old Mrs Fuller was a very dark old woman but, my God, her place was always snow white, everything white about it was white.*

The painful symbolism in this short piece about the daily whitening of the blackened fire by women who kept everything *snow white* even though their skin was *very dark* is revealing about the pressure to conform, even when they were, by definition, unable to meet the requirements. Kathy was old Mrs Fuller's grand-daughter.

I wrote to Kathy in reply and explained that it was fine that she didn't wish to publish any of her stories and that, although she felt ashamed of her 'broken English', I valued it as a way of telling her story that uniquely ex-pressed her sense of the place. I wrote about Patsy's process of coming to terms, independently of me, with her own storytelling style and language in *Ingelba*. I also let her know that I was very happy that she had taken the trouble to write to me and express her opinion. No letter came back in reply and the next time I went to Coonabarabran, I knew we would have to talk it through. The politics of representation in practice. On the next visit Kathy said she wanted to have her stories in the book and we agreed that they would be anonymous. It was not long after that that she took me in when I was sick; she had become my chief support and constructive critic, writing to me frequently to keep me in touch with what was happening.

But now I want to challenge Kathy again; to include her letter with her name attached because it says so much about the negotiation of the form of these stories that I can't say myself. I sip my tea and wait while she reads. Finally she looks up, reading glasses on the tip of her nose and says with author/ity, 'Yes, that's quite right,' agreeing with her words in the letter, 'That's fine,' closes the manuscript and hands it to me.

Each time I go back to Burrabeedee, Forky Mountain and Coonabarabran, I visit Kathy and each time I have to jump through the same hoops. Then I am accepted into the inner sanctum. It's a bit like Kathy said herself when I asked her about the ending of Burrabeedee:

We always keep crossin' tracks with our Burra, we kept goin' back all the time. I still go back out there all the time don't I? I go out all the time, see to the graves and ...

~

That night, home from Coona, I dream again that I am on top of the mountain, but this time it is Forky Mountain. I have large post bags, bags and bags that I have carried up, staggering and stumbling all the way. When I get to the top of the mountain I realise that not only am I at the top but that I can unload all these bags in a little red chapel that is built on top of the mountain. I leave them there with a sense of great and enormous relief and contemplate the climb down the other side.

Performance V

La mer/la mère: and the cassowary women of Mission Beach

Landmarks
Mullaway, *late summer*

At dawn, after a hot and sticky night, I slip on a sarong and walk in cool light air towards the sea, through paperbarks lit pink by low lying sun. On the beach, the sky above the horizon an apricot blush, reflects pink on green waves and lights foam to clear shining white. The sand is flat and smooth from recent high tide, bare except for little crab patterns of tiny pearls in odd petal shapes and fleshy straps of golden kelp that look luscious and good enough to eat.

I swim in clear green rockpools washed by waves, tongue and mouth receiving wetness and salt, holding onto rock ledges, skin responding to pulse of waves, like the opening and closing of blood red anemones clinging in a crack in the rock. I slide a finger into their opening, smooth and slippery as the inside of a mouth; they cling to my finger and suck it in, food from passing waves.

The landscape of this performance is Mullaway Beach, mid north coast of NSW, a place for remembering and production of stories about the cassowary women of Mission Beach in far north Queensland. The beach is a space of pleasure, play and holidays where I explore stories about women's production/ creation in relation to the landscape. I want to find out what makes white women express such a strong relationship to place, how they cross the border from private to public, and where their bodies are located in all of this. Special places

Sunrise, Red Rock, 1997
(Margaret Somerville)

with lots of stories, form knots or conglomerations of energy linked to other places by networks of lines from older and revered women, who like the five matriarchs radiate connections across the landscape. I am interested in taking up Liz Grosz' challenge to put the body at the centre of theorising and extend this idea to place. I work with her ideas about the space between language and the body as she interprets the French feminists, Kristeva and Irigaray. Subjects are seen as connected to, rather than separate from the landscape. The question of body connection is seen as important for all bodies, for the body of the world.

~

> we are ourselves sea, sand, coral, seaweed, beaches, tides, swimmers, children, waves ... More or less wavily sea, earth, sky—what matter would rebuff us? We know how to speak them all (Cixous 1981: 260).

In the beginning there are no words
only the salt water lap lapping at the edge of the sea.

I begin slowly, with rhythms of coming in and drawing away, to have words of my own as I move between shed and water, always drawn back to the water's edge. I am there at dawn, in the heat of the day, at dusk and in the night as I watch the tides grow larger with the pull of the moon towards the summer equinox. At low tide all the creatures who inhabit this intertidal zone with me are stranded on rocks bared to the sun, and at the highest tides the waves cover the rocks where I sit. I am fascinated by all the creatures that live on these margins, opportunists who wait for passing trade. When the tide is out the cungevoi we used to hack for bait are nobbly nodules of barnacles and weeds with holes that squirt as you push them. As a child I was surprised and even dismayed that they were animals when we cut them open to get their meat for bait. Now as the waves lap over them at high tide I can see them all, hordes of them, with their throats open to the waves, delicate triangles of red-pink flesh held open to receive morsels of food that pass over them.

Lillian tells me to write down the things that give me joy: *the sea, music, singing*, particularly women singing (mother, other, m/other). I remember the very sensual pleasure of the deep-throated belly voice of Joan Sutherland trilling up and down my spine with sounds of 'Casta Diva' from *Norma*—the whole surface of my skin luxuriating in this sound. It is as if my body is immersed in watery fluid and the music is vibrating through it as touch. In utero, it is the pure sensation of my mother's voice. My first memory is of my mother singing. Is this what Irigaray means when she talks about multiple female pleasure, when she reclaims Freud's polymorphously perverse sexuality?

> But *woman has sex organs just about everywhere*. She experiences pleasure almost everywhere. Even without speaking of the hysterization of her entire body, one can say that the geography of her pleasure is much more diversified, more multiple in its differences, more complex, more subtle, than is imagined—in an imaginary centred a bit too much on one and the same (Irigaray 1981: 103).

I have been reading Irigaray all day, feeling a little distant and wondering how *the blanket* fits. It just appeared on the surface of consciousness and it comes from getting the knitted woollen squares out of an old pillowslip that I have carted all around the country since time immemorial. They are knitted in garter stitch—mostly in sombre colours of grey-blues and green-blues. I wonder where these colours come from, they are not really my choice of colours, and I decide to knit some more squares in deep reds and maroons, my colours, warm colours—one a night—until I have a blanket. I get out knitting needles and begin. From this beginning the idea grows—the blanket becomes a self portrait, knitting myself together. I want to buy old jumpers from Vinnie's, other people's clothes, and unravel them stitch by stitch, rolling them into delicious balls of crinkled wool and re-knitting them into squares, joining them to the old squares I already have, to make a blanket to throw over the sofa, to wrap myself up in on cold nights, like the words of old stories.

As I touch them, I remember that the old squares were knitted by my maternal grandmother, the Gaelic one, the one that loved me the dearest and who has been dead now for over thirty years, they were her colours—and my mother's. I remember them in their colours, my grandmother was greens and my mother blues, and the texture of their wools. Grandma always knitted in a fine, five-ply Bluebell Crepe which had a double twist and a fine slightly textured finish; my mother preferred an eight-ply Cleckheaton because, like the Bluebell Crepe, it didn't 'pill' but knitted up faster. I sit by the fire and begin knitting my red squares, in-over-through-off (rhythm of sound as my grandma taught me to knit).

> Yes, I am coming back from far, far away. And my crime, at present, is my candour.
> I am no longer the lining of your coat (*I make a blanket/coat of my own*)—your-faithful-understudy. Voicing your joys and sorrows, your fears and resentments. You had fashioned me into a mirror but I have dipped that mirror in the waters of oblivion—that you call life. And farther away than the place where you are beginning to be, I have turned back. I have washed off your masks and make-up, scrubbed away your multicoloured projections and designs, stripped off your veils and wraps that hid the shame of your nudity. I have even had to scrape my woman's flesh clean of the insignia and marks you had etched upon it (Irigaray 1991: 4) (my italics).

I dream a dream in which I am pregnant and am delivered of a beautifully formed creature through a deep blood red slit in my belly. It is a perfect grey-blue sea creature, in between a crustacean and a sea horse, and lies on a bed of red porous sponge-like coral, the lining of my uterus.

This is the chapter I cannot write in the world of Men and Language, so I make plans and preparations to go on a holiday to the sea to remember other seaside holidays and forage in the sand for sea-stories. I ring Kris, scientist/ecologist, the other/mother of this chapter and ask her to accompany me on this journey.

I talk to Laura on the phone about the spiral of excitement that overtakes me as I make preparations for a-holiday a-writing. There is such a proliferation of ideas coming from, and going in, all directions at once, a spilling over, an excess, that I can no longer write it down. It's like being on the edge, the limit of excitement. 'Am I going mad,' I say. We talk about her experience of the excesses of language before a migraine. 'It's like having a foot in both camps,' she says, 'the pleasure is in the unboundedness of it, in being able to access both the semiotic (body/mother) and the symbolic (language/father), to draw on both simultaneously, without conflict.' She says she wants to celebrate exactly the opposite of unhingedness; it is taking the thetic[1] possibility to its absolute limit with its structures and connections that gives sanity and meaning. It is the opposite of unhinged, it is to be there and to be absolutely sane that makes the experience so precious.

> The semiotic is understood by Kristeva as a pre-oedipal, maternal space and energy subordinated to the law-like functioning of the symbolic but, at times, breaching the boundaries of the symbolic in privileged moments of social transgression (Grosz 1989: xxi).

I am interested in exploring the residues of connection/fusion to the m/other I say to Laura, 'the material order of the textual/psychical trace or imprint' (Grosz 1989: 42). I talk about the blanket and how a blanket is often the first object of desire that makes the transition between child, mother and outside world. It has its existence in the space between self and other, self and world. There seems to be a connection through the blanket between the body and the abject, that the blanket is made from the abject, the extrusions, the products

---

[1] The thetic is 'the threshold between the semiotic and the symbolic ... as well as the residues of the semiotic in the symbolic' (Grosz 1989: 45).

of the body that are neither inside nor outside, but both inside and out.

> The abject cannot be readily classified, for it is necessarily ambiguous, undecidedly inside and outside (like the skin of milk), dead and alive (like the corpse), autonomous and engulfing (like infection and pollution). It disturbs identity, system and order, respecting no definite positions, rules, boundaries or limits. It is the body's acknowledgement that its boundaries and limits are the effects of desire, not nature (Grosz 1989: 74).

Laura says it is important that this object is irreducibly grounded and material, even more than the body. How can something be more grounded and material than the body? Its central quality is that it is able to be perceived by the senses—smell, taste, touch, colour—and its enveloping potential. I think of the blankets my babies had, the frayed end of ribbon edging rubbed against skin, the sooky smell from wetness of mouth, its decaying potential just before it finally went into the bin.

> Her [Kristeva's] notion of abjection sketches the peculiar space and time (given that we cannot yet talk of a subject) marking the threshold of language and a stable enunciative position ... Understanding abjection involves examining the ways in which the inside and the outside of the body are constituted, *the spaces between the self and other* (Grosz 1989: 71) (my italics).

Kristeva's abject is always a space of disgust. I want to turn her abject on its head and claim as a positivity this space of becoming, the not quite separate, not quite the same, the space of the abject, which allows one to be *corporeal* as well as *speaking* subject, out of the fluid space of the sea.

That night I begin my red knitting from the unravelled wool of other people's clothes (Lacan's 'body in bits and pieces')[2] when I can't sleep. As I knit, I think of the photograph of the women in the local paper knitting squares to

---

2 See Grosz (1990).

stitch together to make wraps for refugee women in Somalia, another reaching out, crossing boundaries. I will do that later, right now I will knit a blanket to take with me.

~

The morning before I leave for the sea I dream of my mother, as she was before she died, my eldest daughter as she is now, a young woman, and myself, with some shadowy male figures, interchangeable sons husbands fathers. In this dream my mother wants to go to London and has not asked my father. I am angry with her and tell her that it is all right to want to go to London, that *she must speak her desire*. My mother is quietly insistent that although she has not spoken it, she is going anyway. I am also angry with my daughter who will not ask for what she wants but in this dream I know that she is young and has a long while to learn.

The material of this dream is images and facial expressions, psychic connections of three generations of women. It is an immediate and powerful experience and on waking it is as if my mother is with me and speaks to me. I had always explained her death by cancer with the story that she didn't express herself, that she gave up on her creativity, her desire, and it turned inwards as a cancer and consumed her. In this moment I recast her story—she did, in fact, 'go to London'. After my mother had the first sign of cancer she was presented with a choice, either she could go to Oberammagau to see the Passion Plays which she had wanted to see since she was a young girl—or she could stay and receive treatment for her cancer. Mum chose to go to Oberammagau with her best friend. She risked her life in that choice, possibly even died because of it, but she made it very clear to me, in this dream connection, that it was what she chose to do, to live her desire until she died; to even choose her death. This profound permission-giving reversal of storyline was made possible by my immersion in the sea of this new story of mine.

> I wished that that woman would write and proclaim ... I, too, overflow; my desires have invented new desires, my body knows unheard-of songs. Time and again I, too, have felt so full of luminous torrents that I could burst—burst forth with forms much more beautiful ... And I, too, said nothing, showed nothing; I didn't open my mouth, I didn't repaint my half of the world. Where is the ebullient, infinite woman who ... feeling a funny desire stirring inside her (to sing, to write, to dare to speak, in short, to bring out something new), hasn't thought she was sick? Well her shameful sickness is that she resists death, that she makes trouble (Cixous 1981: 246).

Walking with Mary on waking I reflect on the name *Mission Beach*, it is strange that I have never known what it means. There has always been an absence of the Aboriginal in my work with the cassowary women while so much of my work has been about old missions. What does Mission Beach mean here?

Mission—meaning evangelical religion, enclosure, repression, cultural erasure—or *mission*—meaning journey, quest? And *beach*, that marginal space between sea and land, that golden-warm site of our pleasure and play. Perhaps I need write no more than the name, I say, the tension between these two words says it all.

It is nearly time to go and I say goodbye to my filly foal daughter on her way to dance camp, sitting smugly in the train, cheeks bright with excitement, ears plugged in to the music of her dance, already my goodbye is erased.

Mullaway, on the first day

> And God said let there be light: and there was light.

I pack the car, pick up Kris and slide down the mountain past the first distant sight of the sea, where the elements of earth, sea and sky merge in a blueness of bliss.

> We're all going to the seaside,
> We're all going to the sea,
> We're all going to take it easy

Where the air is bright and breezy,
We're all going to breathe the ozone,
And be as happy as can be.
We're having a fortnight out of town,
Coming back all nice and brown,
We're all going to the sea.
(Mother—song for journeys to the beach)

We pass the journey by telling stories.[3]

Tell me the story of how Mission Beach got its name?

There was an Aboriginal Mission Station up the back of South Mission Beach on the road between South Mission and Tully. The story as I've been told is that the Aboriginal people were rounded up and brought to live at the Mission. It seems that it was the people of the coast.

Sometime early this century, about 1918, there was a huge cyclone which destroyed the Mission. Accounts say that the Superintendent and his daughter were both killed by flying sticks but I don't recall people saying about what happened to the Aboriginal people or how many died. A lot must have died because the Mission was devastated. After the cyclone, the people that were left were rounded up and shifted to Palm Island.

They say that all the sand that makes Mission Beach came with the cyclone, was dumped there, which is where the white people live now (laughs). What the Lord giveth, the Lord taketh away.

We arrive at Mullaway in the cold light of that winter afternoon and eat salty fish and chips by a dark ultramarine blue sea while the blue islands offshore flicker in and out of visibility. I am wondering what these islands are that come and go and what is the connection between this cold blue coast and the warm aquamarine water of that other undeniably tropical journey to Mission Beach in far north

---

[3] I am indebted to Dr Kristine Plowman, for her generous sharing of personal stories and stories of science, an essential part of this performance of 'environment' as place.

Queensland when a nearby sign seems to respond to my musings:

> The Solitary Islands Marine Reserve where warm and cold currents mix to produce a fascinating variety of marine communities. Marine life from the Great Barrier Reef and as far south as Tasmania can be found living together here.
>
> The Reserve protects aquatic plants and animals by safeguarding natural habitats that include islands, submerged reefs, headlands, beaches and estuaries.

The sea will carry us where it will. Cyclones shifting beaches, islands coming and going, currents transporting marine life from the tropics to temperate waters, there is movement, change, unsettlement. 'The ocean is the supreme metaphor for change. I expect the unexpected but am never fully prepared' (Winton 1993: 40). I watch the water and contemplate my longing to be immersed in the cold blue waves of the sea.

Tell me the story, I say to Kris, of how we come from the sea.

I see the sea as like our evolutionary past, it's where life began. You think about the amniotic fluid, we actually have a little sea inside ourselves for our babies. The fluid in our bodies is made of the same stuff, it's nearly the same ionic balance, and it's full of salts and I think it calls us back. I think we are of the sea. I think that's part of the sea's attraction to us, that ancient knowledge that we carry in us, we can't survive in it that way any more but we have evolved from it, and we are of it.

When we are in utero we go through a phase where we are sea creatures, with gills. The stapes and anvil, the bones in our ears, were the gill struts and jaw bones of our ancient fishy relations. In the fish there is a thing called the lateral line which runs around the fish and this picks up sensory information; the ear develops in other vertebrates like us by that lateral line forming a pocket and then, as that pocket becomes more elaborate, the bones of the ear pick up sound vibrations.

People interpret these things in different ways, of course, but it is as if we repeat, in some sense, our evolutionary history in our development. We evolved from the sea to the land and the sea stays with us right through the vertebrates in the amniotic fluid, that archetypal fluid that we maintain to overcome the fact that we are not in the water any more. It's like we say our history is written on our body, and we just mean our contemporary history, and so it is, but in fact, all our history is in our body. The very beginning of life is in us, each one of us carries the story of life on this planet in ourselves, and we add layers to it and we shift it, we are as old as the planet and we are made of the pieces of it.

I incorporate the 'stories' of science into the fluid space of the sea where they are just one of many stories thrown up onto the beach.

That night, just as the light fades from the sky, we swim in flickering white foam on the edge of cold waves that take our breath away and then bring the blood rushing to the surface; a tingling all over, a hot shower and rug up for the night, read and tell stories by the fire. We read aloud from Irigaray—from *Marine Lover*—for whom the sea represents the space of the semiotic, of a female life force, an idea which appeals to Grosz:

> The marine element is therefore both the amniotic fluid, the deepest marine element ... is also, it seems to me, something that represents female *jouissance* quite well, including in a movement of the sea, of going and returning, of continuous flux which seems to me to be quite close to my *jouissance* as a woman (Irigaray 1981: 48–49, cited by Grosz 1989: 170).

Can I write female jouissance in the rhythm of the waves? Is there a way of writing the landscape that goes beyond the dualisms inherent in our language?

We begin by telling stories from the seaside.

∼

My first journey, just a few years ago, to Midyim, Kathleen McArthur's beach house at Caloundra, hidden by twisted paperbarks, coastal tallowwoods, pandanus, hanging with vines of native wisteria. Downstairs, cool and spartan with concrete floor, beachshower-bathroom and guest room, then up cool and dark varnished timber staircase to the smoothly polished, soft-as-satin, crows ash floor of the living room. The room is lined with hoop pine and built-in timber benches and opens to the outside with two windows that slide out of sight. Two tables and rows of bookshelves hold Kath's life's work. We sit and talk at the table by the open window that lets in breezes and sounds of the sea.

At night we sleep in a tiny dark timber-panelled bedroom with four hard handmade bunk beds, like the beds of my childhood beach holidays. There are two small windows where a light breeze stirs the leaves and the moon shines on twisted trunks. In the morning Kath shows us a cluster of golden brown native orchids amidst the dark and tangled mass of her garden. And downstairs, behind the guest room and bathroom, the secret place of her wildflower paintings, a gallery still hung with masses of her beautiful wildflowers. She gives me a pack of her wildflower print cards. For years Kathleen has sold her paintings and prints, not only to make visible her local wildflowers but to raise money for her campaigns. She talks about her nursery where she has long grown and sold the same native plants, the love of her life.

Later, in exchanges of letters and books, Kathleen sends me six volumes which document an extraordinary and richly productive life, beginning with her beachside haven at Caloundra:

> Caloundra, where winds gave each day its special character; where there were uncrowded beaches ... where rockpools were full of fascinating creatures that went about their living undeterred by our peering eyes; those sunrises and moonrises backlighting the cloud formations on the eastern horizon and kaleidoscopic sunsets over the Glasshouses; with whales playing in the bay unbothered by the Port of Brisbane shipping; with Pumicestone Passage full of oysters; black swans flying over at the end of the day; and wildflowers, wildflowers everywhere ...

They hugged the windy headlands, competed for space in the maidenhair gullies, carpeted the ground under the paperbarks and positively rioted over the pademelon plains (McArthur n.d.: 11).

And Kath painted them all. It was not only an act of desire, of love and of female productivity, but a strong political gesture, both in the movement of the private female aesthetic into the public realm and in her increasing involvement in political campaigns to save the places where the wildflowers grew. What began with her garden and love of the beach and wildflowers, led to the establishment of the Wildlife Preservation Society of Queensland (WPSQ), a network of conservationists still active throughout Queensland. Kath shows us new ways of seeing the beach and the coastal hinterland.

~

Mullaway, on the second day

And God made the firmament and divided the waters which were under the firmament from the waters which were above the firmament: and it was so.

It is cold and and wet as we struggle across grassy headland into wind. Walking in this harsh sea wind and rain I have to meet the energy of the elements, push hard against it, walk fast until surge of blood warms face. Fingers and toes tingle as ideas flood to surfaces of body. My whole body moves with growing ideas, whole body involved in their becoming, the energy of their production. How to represent the richness of associations, images, patterns, ideas and textures, being in the landscape and its stories?

On the lee side of the headland the plants grow tall and free, great banksia trees with burnished gold brushes like ritual candles studding their dark trunks and casuarinas, gnarled and ancient, with sparse silver needles dropping in oriental shapes at the sheltered end of the beach.

As we push against the force of the wind, the plants are more stunted and flat to the earth, old gnarled trunks running almost horizontal, clinging to the shape of the headland. We learn to move with the rhythm of wind and sea.

Only the pandanus stands free, bare trunked with palm fronds and massive buttress roots holding them fast against wind and sea, growing in the harshest of conditions on the exposed face of the headland. *Pandanus forsythii*, we call them; Aborigines ate the crisp and succulent tip of the prop roots and the inner growing tip of their leaves. At the base of the headland, sheltered from rain and driving wind, we poke about in multicoloured rockpools and one of the colours becomes a tentacle, long and curling, in and out of the shape of the rock. Looking closer into the darkness there is a bulbous body/head with two stalky eyes looking from either side of its mottled head mound. Tell me the story of the octopus I say.

Octopus are like this (hand splayed open) with all their tentacles; there are eight tentacles, and they do elaborate courtship rituals; they have a package of sperm, like an envelope that keeps it from being dispersed in the water; she pops it into a particular place and opens it up when she is ready. Lots of invertebrates do that, the male and female exchange packets of sperm, and sometimes it has to last her entire life which might be several broods of young she produces, and she has the means to preserve and store those sperm until it is needed for her ovary.

When you eat them you have to take out the ink sack, and the gut. But it's the ink sack that doesn't taste very nice. And in the squid you take out the cuttle, a bit of cartilage, the strong thing that those muscles contract against. And you skin them, they've got that wonderful skin, its soft, a bit like ours, and its easier to cook without the skin and easier to eat when you skin it.

That night we eat squid and baby octopus and little leather jacket fillets cooked in ginger and honey and tamari and read stories from the Mission Beach journal.

~

North Queensland, 22nd October, 1991

Tomorrow museum researchers will collect coral spawn. At three to four days after the full moon the coral releases millions of gametes into the sea, making the water thick and soupy. The shelves are full of rows and rows of white lacy coral in its multitudinous forms, all with little animal holes where once the coral animals were still alive.

North Queensland, 23rd October 1991

This morning, swimming early at Mission Beach, the warm flat sea is crossed by swirls of golden-brown moving freely with current and tide. Closer, they look like fine grains of sand but they're floating thickly on the surface of the sea in random swirling patterns. It is three to four days after the full moon—and we swim in the fecundity of coral spawn which leaves no apparent trace on the skin, has no obvious taste and no sensation to the touch. The water is warm and we float for a long time, moving gently with the coral spawn on the currents. Later, in the shower, the coral spawn has left traces of slightly oily golden-brown swirls across my breasts like the patterns the women painted with kangaroo fat and ochres in the desert.

~

Tell me the story of the coral spawn.

At very special times, the corals from the outer reefs, I think it is October, on the full moon, release their sperm and ova, and the inner corals do it on the full moon in November. The moon is the synchronicity, the moon lets them know it's time—so out they go. The sperm get released first and then the ova and they fuse in the water to form a new individual and it floats around in the sea for a long time, it can go quite long distances and it will settle down on a surface, if it finds the right surface, and turn into a coral. Millions and millions of these are released and only very few ever become settled and of those only very few become a coral. It's the currents that take them wherever they go; the currents are moving around and they're just flotsam and jetsam floating with the currents, and some of them are taken really long ways. It's sort of like a huge abundance of life that spills out, all together, they've got to go out all together, cause the sperm's got to run into the ova, and a lot of invertebrates in the sea do that, they have times like the full moon, times when its warmer, or it may be the tide, the high tide too, all these gametes are let go to the adventure of life. Those little invertebrates produce lots and lots and lots of gametes, it's seen as expensive, that's what they say, excessive.

I have been 'written on the body' by the landscape, by the excesses of coral spawn floating in the full moon.

Mullaway, on the third day

And God said let the earth bring forth grass, the herb yielding seed and the fruit tree yielding fruit after his kind whose seed is in itself, upon the earth: and it was so.

We go exploring a little way up the coast to Red Rock. The beach is so rich in colour and form that it seems to invite only a short ritual presence. At the southern end, just south of the confluence of river and sea, is the red rock, liver-coloured bodily extrusion, as dramatic in form as colour, marking this sacred place where river meets sea. What Aboriginal stories about this red rock are washing in these waves? On one side of the rock a small sheltered beach, with silver grey casuarina leaves dripping towards silver white sand, curves in a quarter moon to a jutting headland. On the other, the bank of the river, extruded red and white, curls and crumbles with its fragile colonies of banksias, casuarinas and pandanus.

We walk around the curve of the river over pebbles and oyster shells; purple lilli pillis grow on the river bank, and someone has just caught a flat silver moon fish. I forage for stories about foraging for food.

Well, you just walk around the rocks and look for them—there are different sorts of oysters, there are little ones that are really hard work so you tend to leave them till last—and you have an oyster knife, they're flat with a point at the end, you just prise the top off the oyster and after a while you learn where to put the blade and it can cut back onto the abductor muscle, so you slip it in and slide it that way, and just take the lid off, you can do it so that you just slide the oyster out and then you pop it straight in your mouth, and there's something really nice about struggling over the rocks and collecting and eating them when they're fresh. The eugari, you can eat them too—like the pippies, they get sand in them so you put them in a bucket of salt water and because they open and shut and open and shut, they wash themselves out.

And that blue lilli pilli?

You can eat all the lilli pillis, they're all really high in vitamins, they can be very tart, they make a nice—what do you call that thing—sorbet—some people use them in the really flash places as a palate cleanser between courses—and you can get lilli pilli ice-cream or you can get sorbet.

That night we forage for stories in the body/landscape journal.

~

North Queensland, 23rd October, 1991

At a little wooden cottage on a bare hillside at Mission Beach we meet Karen and Robyn who are on duty for the Movement for Responsible Coastal Development and chat to them over a cup of tea. Karen joined the local 'progress association' and it was full of antiseptic people 'who worried all the time about cleaning the mould in the drain'. 'Why,' I asked her 'aren't you worried about the mould in the drain?' 'Because my mother was a conservationist,' she said, 'even my father was interested in conservation.' I asked Rob why she wasn't worried about 'the mould in the drain'. 'I had four children and didn't have time to worry about the mould in the drain, I am more worried about the world I leave behind for them.'

We talk about the cottage and the cassowary project and institutionalisation. They have decided to go subterranean again, meet in someone's garage one day a week. We probably won't get much done but it will be a lot of fun. 'If you don't watch out you'll be out on the streets,' Kris says and Rob raises her eyebrows, 'It might be fun on the streets.'

A subversive project of fun and pleasure. Robyn and Karen are working on the design and layout of a calendar with a big picture of the cassowary, strutting brilliant blue plumage and red wattles, looking the viewer straight in the eye. There are smaller pictures of the chicks in a month by month story of their development, and we juggle photos to select the strongest visual impact. The visibility of the bird is important in getting public support. We swap ideas about sewing pants when I admire Rob's big billowing light and dark green patterned pants, worn with a loose white T-shirt, purple patterned cummerbund, soft grey hair in a loose top knot. Batik on calico, she loves calico; because it is still strong enough, once treated and boiled, waxed and dyed, to make into pants, a simple triangular pattern with no side seams, adjusted for crutch to belly button size. She will send me a pattern. We leave with bananas and sweet bite tomatoes from a gift box at the door.

～

The cassowary project was predicated on the visibility of this normally invisible bird. It appeared on Mission Beach after the 1980s cyclone when all its food plants were destroyed and it was forced out of the rainforest to forage—and people thought it was beautiful.

Tell me the story of the cassowary.

It is beautiful, it's big, it's dramatic, it's mysterious—it's all those things—and it was vulnerable. Before the cyclone they just did their own life, but when the cyclone came it was like people had them, they were vulnerable, they fed them, some people had their whole identity tied up with feeding the cassowary. When the cassowary became visible, all that development stuff was going on and the department was going to release more and more Crown Land. What the cassowaries did by appearing was to present people with the notion that this was a vulnerable thing, this forest, and that when the next cyclone comes there might be such a small block of rainforest left that it will be erased forever.

Before the cyclone when you drove to Mission Beach along the dirt road, the canopy of the rainforest met overhead and it was a little dirt track through the jungle that led to a beautiful beach. After the cyclone the Main Roads thought it was too dangerous so a huge swathe of rainforest was cleared. Mission Beach had previously only been a holiday place for people from Innisfail and Tully, but when the road was upgraded, the tourists started to come. Then, in the boom years of the 70s and 80s, everybody was going to develop everything—huge marinas and resorts were planned. The Lands Department and Cardwell Shire Council began dividing up and selling off Crown Land and a group of people formed the Movement for Responsible Coastal Development and said 'No'. In the end the cassowary became a powerful symbol, an icon for North Queensland and the rainforest; it was pushed to be a powerful symbol because of the cyclone.

Mullaway, on the fourth day

> And God said let there be light in the firmament of the heaven to divide the day from the night; and let them be for signs and for seasons, and for days and for years.

We walk to the nearby beach village of Woolgoolga, visible after we cross the first headland, white houses and a Sikh temple nestled into bare grassy headland, like a fishing village on the Sea of Galilee, a pleasure place for peasants, *fishers of men*. Closer up, the beach is bordered by park and caravans taking advantage of beach and sea. One has a Greek garden in polystyrene boxes—leeks and ochra, beans, tomatoes and chillis, spinach and capsicum—and in the front, chrysanthemums and herbs.

On our way back I turn for a last look at the little town. Clouds have come up, a deep indigo, and the sea emanates its own light, a clear translucent green. At the far end of the long straight beach, above the little village on the headland, the sky opens up to a light green vortex echoing the translucent light of the sea. The light shining from this opening reflects on the little houses and the temple as if a vision is about to appear.

I recall the visionary excesses of Revelations and images of the second coming of Christ. I wonder about the sea as site of visions and stories of the spirit.

Our Christian stories, Jewish in origin, give us a spirituality of denial, denial of the body and conquest of the landscape. Can our stories of pleasure and bodies, and the landscape of our sea, give rise to a different spirituality? What do we have of an indigenous spirituality, a ritual of landscape from the land in which we belong? What of a spirituality that includes women, that does not exclude the body, that connects us to this place in which we find ourselves?

In the *ngintaka* story Nganyinytja tells us about the creation of a landscape.

Wherever he went that explains that geographical phenomenon, and there's a salt lake over there where he scraped his tail. At Tjanmata a nice beard came forth and then he went out in that direction and you can see where he perfected and placed the rocks at Arting and wherever he stayed usually a mountain resulted. Where he vomited, you will see what he vomited. And wherever he stopped and re-gurgitated he produced a whole lot of ngintakas. *And what he brought up is now the various foods that they get in this area (kaltu kaltu). Kaltu kaltu is one of the best seeds that they grind to make a cake. And when his stomach got rid of everything it was empty and it was a big cave and there's a big cave you can see there. So this whole area is called* ngintaka *country.*

At the re-gurgitation site, Nganyinytja describes how all the *mayi*, the plant foods, were created from the abject, the vomit of the *ngintaka*.

> *This is the back of the* ngintaka.
> *These marks represent the seeds that he brought up.*
> *And he ground up the seeds and peopled this land here with the plants that we eat.*
> *Before the rains came he put out all the* unimpa, *seeds.*
> *He ground them up*
> *and these too, he vomited them up [the rocks].*

In ceremony the women re-enact this creation story, the making of the grass seeds from the abject.

~

I searched out the story of the cassowary women a year after I first met them, when Rob sent me the pattern for the pants and I wrote to ask whether I could come and talk with them. I asked how they came to Mission Beach, how they got involved in conservation and about the cassowary project. Karen:

As a child, my first memories of it were coming through the dirt road. It would have been in the sixties, and I can remember it always being a super long trip and really lovely dark jungle that we went through, that dirt road was completely jungle all the way, and there was a little beach at Innisfail called Ettie Bay which was always a favourite. And I can remember, we would come down here with my grandmother or aunt and there was always this thing between the adults about whether Bingil Bay was a nicer beach than Ettie Bay.

Childhood holidays and pleasure. Norma:

We used to go to Hervey Bay for holidays but one year we said we'll go all the way to Cairns and someone said to call in at Mission Beach so we called in there and we never got to Cairns. We stayed there each year for the next couple of years and we went back to South Mission for our holidays 'cause the fish and that was incredible. Well, one trip we arrived and you used to be able to just camp on the beach front, this lovely beach and Dunk Island and all the other islands, and the purple water and the blue sky—and one year we came back and this camping ground was going up. We were sitting there one morning and I said to Bill, there's only one answer to all this, we'll have to get our own block of land that we can put our caravan on.

Holidays, pleasure, unsettlement and change.

Norma sailed to Mission Beach from Sydney with her husband when they came to live in 1979. She noticed the changes in river and sea as they moved their yacht from mooring to mooring and sailed through the channels in the river mouth.

But what seemed to have happened over the years, and this gives me great pessimism for the reef, is that with the rush of development and the expansion of the sugar and banana industries, they're all clearing the rivers and there's an enormous amount of siltation taking place and we've seen it because when we used to come in the seventies, the coral was beautiful, the water was clear—like I said, the first time I walked on the beach the water was almost not there it was so clear. But now you go out on the reef it's covered in fine brown silt, there's no coloured coral. What little bits of coral you see poking through the mud is just the tough white corals, there's no soft corals and this really makes me worry for the reef itself. We found in the yacht that, in the beginning, we could go in and out of the river entrance in the pitch dark but with all this silt coming down the mouth kept changing and eventually it got so silted up that it wasn't deep enough.

Taking pleasure in the landscape led to their knowing and loving their place. They noticed little changes in the body of their place.

The cassowary came to symbolise the body of the place they loved. When I asked Robyn what the project meant to her she said:

Save the cassowary and you save Mission Beach (laughs). I guess it would be very hard to explain to grandchildren that you lived in an era when the cassowary became extinct, that would be terrible. When you had the knowledge to know that you shouldn't let these things happen. I would hate to be part of an area that knowingly let something as majestic as that—any form of life really—become extinct because of what we did. But in the meantime, as well as saving the cassowary, you save an incredible amount of other things because of its size and range.

I remember the cassowary from younger than ten probably, having come across a cassowary in the wild. I've always thought of them as something special, and I've always associated Mission Beach, the Bingil Bay area with cassowaries—the cassowary is very important but maybe it's a symbol of what it's all about too. If the cassowary had never lived here I would still want to save as much of this as we can 'cause it's nature, it's beautiful.

The body of the cassowary is only part of a complex system of bodies.

The cassowary is critical to the wet tropics because the cassowary is the only bird that can revegetate rainforest with trees of certain types, the only bird big enough to distribute. If the cassowary goes the world rainforest will change, there will not be the distribution of certain seeds that only the cassowary distributes.

And the body of the cassowary produces the seeds that grow the rainforest that feeds the cassowary that produces the seeds ...

It started off with a little bit of a nursery at Margaret Thorsborne's place at Cardwell and she used to hand trees around to schools and then we had that vacant block of land over the opposite side of the road so we offered that to the Wildlife for the nursery. We collect seeds from cassowary droppings so we know they're food trees. Its incredibly creative—I don't know that creative is the word, but satisfying probably. You collect the seeds and then put them into boxes and then the little seedlings pop up, and once a month we have a potting day and we probably bag seven or eight hundred trees in an afternoon.

And the body of the cassowary is part of a complex system of bodies that are inextricably intermeshed in their living and their survival.

The cassowary has to have an area that has a variety; they need to have fairly complex vegetation because they can't just feed on one tree because it doesn't produce seeds all year round. And this is the reason why Mission Beach has the population—because we've got the variety of vegetation.

The birds—like people think they're rainforest but they're not explicitly rainforest—they come down into the melaleuca forest, the lowlands rainforest, and into the mangrove swamps; they have to follow the fruiting of the trees so that is why they need such big habitats.

And the body of the women is the same body as the body of the cassowary.

Joan's got this particular bird and his name's Delinius. She habituated herself with Delinius within about six months of finding him, he accepted her and she became part of his routine. He actually took her and showed her his nest which is absolutely rare and he sat for forty-eight days without eating or drinking, never left the nest. When the chicks are born, she never speaks or touches them, just follows them. When he sits, she'll sit, whatever he does, she does. So with the chicks he sat down, the little chicks were growing up and she said, whenever I sit down, the chicks always have to come over and interfere with me. They peck her things, they peck her watch, they look in her bag or all this sort of thing, they just can't keep away from her. Anyhow Delinius got up to leave and Joan was still working and didn't get up. The chicks sort of went when he went and they turned around to Joan as if to say get up, and they sort of walked backwards and forwards as if to say well he's gone, aren't you coming too?

~

In that undefined space of connection between self and m/other the cassowary women grow forests from the abject in a different creation story. My question then is, how can we change the dominant storyline of landscape representation so that this is the story that is inscribed on the landscape rather than the story of the developers?

Mullaway, on the fifth day

> And God created great whales and every living creature that moveth, which the waters brought forth abundantly, after their kind and every winged fowl after his kind: and God saw that it was good.

We explore Station Creek, a National Parks beach a few kilometres north of Red Rock. A long drive off highway and then a long winding walk through silver moon craters and rolling dune hills. Fine white sand is dotted with tufts of silver grey spinifex and a purple flowered creeper that spreads fingers across shifting sands. From the top of sandhills a long stretch of undeveloped ocean beach opens out, fine white sand marked by wind, waves or dots of recent rain. As we walk softly, the only other pattern is a fine delicate birdtrack that marks the sand with a single clear inscription, and on the wet sand patterns of dots left by invisible sand crabs.

Then suddenly, unexpectedly, the sand is churned by wheel tracks, ploughing up and down in a highway-width channel through the middle of the beach.

I walk to the end of the beach to settle jarring cells in the quiet of rockpools, but I pass four-wheel drive picnickers sitting in the back of a truck on the beach. Wave upon wave of anger immerses me, I am churned up like the sand, and want to shout and roar at them, reveal their offence. We climb a rounded knoll, a shell midden, thousands of years of eating shellfish overlooking the sea. On the other side of the knoll, through a huge spreading pandanus tree, a smaller beach, made not of sand but pebbles, a million different sizes, shapes and colours—no four-wheel drive marks here—the pebbles move and roll with the waves. As I fossick through piles of sea sponges, driftwood, sea shells and seaweed, amazed at this mass of form and shape, I wonder how the haphazard multiplicity of the sea can answer the single-minded aggressive domination of the tracks?

> In demonstrating that there are other possibilities (of sexuality/textuality or pleasure/production), Irigaray makes clear the violent appropriation by masculine representational and libidinal economies of a richly heterogeneous field of possibilities. *A language that is isomorphic with an autonomous, non-reductive femininity and pleasure would have to overcome the domination and universalisation of the masculine* (Grosz 1989: 130) (my italics).

I fossick for words in the beach book (Drewe 1993) to help me understand.

Drewe (1993: 1) writes that 'although Australians have been conducting a life-long love affair with the beach', there is little literature about the beach in Australia. He quotes Swinburne's love affair with the sea: 'My lips will feast on the foam of thy lips ... /Thy sweet hard kisses are strong like wine/ Thy large embraces are keen like pain.'

To me this sounds like another slightly more acceptable way to bend the sea towards the desires of men. Other stories in the beach book all reflect this one way passage of desire from man/woman to sea/landscape with one exception: a story by New Zealand Maori writer, Keri Hulme, who portrays the sea as body, a force to be reckoned with, a source of food and nourishment and also of death and decay. For her, the sea will throw up what it will, it has its own desire, it is the *other* and the movement in her story is between a self and another which is the sea. She writes of her friend, killed by the thrashing of a sea lion against his boat: 'I gave you to the sea. I rolled you down the sloping floor onto a quarter-moon of gravel and let the seas take you' (Hulme 1993: 202).

The sea is not an other, not to be conquered and exploited but to be listened to, respected, celebrated and feared, just as Probyn (1993b) (Performance 11) writes of crossing over to the space of the other with empathy, imagination, and limit-attitude. The movement between self and landscape as other: it is the same as this movement between self and other as person—across a space of difference.

We talk about the meaning of the tyre marks on the beach. The car as extension of body, body as power and eros, the rape of the land. And yet these marks can be read as just another story among the flotsam and jetsam that the sea throws up, able to be accommodated and inhabited in the same way. Creating a new relation to this landscape may be not about prescription and control but about making visible different marks, different stories. Women have always created themselves in some circumstances, and under some conditions, as desiring subjects but that creation may not have become visible.

The cassowary women create themselves as desiring subjects who have moved beyond the confines of domesticity into the public realm and have thereby made a different story visible. But the danger of visibility is re-appropriation by dominant discourses, and even for the cassowary women it seems that only while they remain subterranean and subversive, will their desires not be subjugated to the dominant storyline. The challenge is for women, through a process of continual displacement, to continue to create themselves as desiring subjects in the public domain, thereby producing a different reality.

Here I come to Liz Grosz's question about refiguring female desire (in the world of received knowledge) as production rather than lack: 'Can desire, in other words, be re-figured in terms of surfaces, and surface effects, to re-make what is attributed a passivity into an activity?' (Grosz 1992b). She looks to Spinoza, drawing on his ideas about desire:

> Desire is a force of positive production; the energy which creates things and which forges interconnections between things, something inherently unstable and changing, so it is not a question of being, attaining a status as a thing, a permanent fixture, but

of ever moving and changing, being swept into a multiplicity of flows (Grosz, 1993).[4]

In this moving, changing world of pleasure, sea and sand, it is easy to imagine such shifts as we might need to re-write ourselves as desiring subjects. The holiday is a space of play, the beach is a space of margins, of liminality, a space of becoming in which change can be sustained as a continuing displacement of a fixed and stable self.

Mullaway, on the sixth day

> So God created man in his own image, in the image of God created he him; male and female created he them.

We find a spot at the end of the closest beach and shelter in a body-shaped crevice at the base of the headland. At this sheltered end, pebbles and shells are scattered by waves in pattern after pattern. I think of scattering words from my fingertips with the same flowing ease. The rock is worn into holes by wind and sea, and lying face down I see whole worlds in these hollows. A tiny fist-sized orb of golds and silvers, purples and pinks, worn to a smoothness, not much bigger than grains of sand. Then I lie on my back, sunhat over face, lost in patterns of sunbeams and smell of sunscreen, salt and straw: Virginia Woolf's (1976: 65) world from the inside of a grape.

> Write! And your self-seeking text will know itself better than flesh and blood rising, insurrectionary dough kneading itself, with sonorous, perfumed ingredients, a lively combination of flying colour, leaves and rivers plunging into the sea we feed ... our seas are what we make of them, full of fish or not, opaque or transparent, red or black, high or smooth, narrow or bankless; we are ourselves sea, sand, coral, seaweed, beaches, tides,

---

4 These extracts are from an untranscribed tape of a paper presented at the annual Australian Womens Studies Association Conference and hence there are no page numbers.

swimmers, children, waves ... More or less wavily sea, earth, sky—what matter would rebuff us? We know how to speak them all (Cixous 1981: 260).

On the last night by the sea we read from our meeting with Margaret Thorsborne, a slight, birdlike figure in a fairytale house in a fairytale world, a perfect setting for re-creating myths of landscape.

~

North Queensland, 29th October, 1991

Huge orange trunked paperbarks lean over narrow dirt track and scrape the car as we drive to Margaret's cottage deep in lowlands rainforest. A wheelbarrow and shovel stand against a tree in an open space where Margaret is planting. Then a clearing with tiny white weatherboard cottage, green shutters and a wallaby who stands up to watch us and then hops off into bushes. Margaret appears in apple green sweatshirt with cut off sleeves and baggy cotton knit trousers, brushing back loose strands of silver from her face.

At the back the house is all verandahs opening into rainforest. There are big scarlet flowers on chintz bedspread; embroidered guest towels in bathroom, and verandah boards are bright yellow with red flowered deck chairs. Margaret wears the pink canvas shoes on thin bird legs like the birds that come to drink. The dark leafiness of rainforest grows up to verandahs where we sit and talk quietly as small birds flutter in shallow water of upturned garbage lid. 'Pitta pitta,' she croons softly as a brown, ground-hopping bird appears through the bushes. 'It makes a pitta pitta sound', she says, and laughs as she tells how, when she returned from a recent visit to Cairns, she called and called to the pitta pitta bird where it usually comes from, then realised he was already there, standing beside her on the verandah.

She leans forward and tells a story about Joe Galeano. He is a wonderful man, his parents migrated from Sicily and he grew up wild on the nearby Hull River. He loves crocodiles and is worried about the silting up of the river mouth which will stop the crocodiles moving in and out through the estuary. He took some people down the Hull River looking for a little sick crocodile (the little ones hang suspended in the water vertically from the surface). They found the little crocodile and a fellow who was with them picked it up, it was very small (as big as a hand), and he cradled it in his old horny hands. 'Be careful, be careful,' Joe jumped up, alarmed, she said, not because the crocodile might bite the man, but because if it did, the baby crocodile would hurt its teeth on the man's leathery old hands!

She feeds us moist apple tea cake with slabs of butter, and tea in delicate china cups. As we talk she points out two green tree frogs who currently live on the hanging basket of brightly coloured fruit, and explains the difficulty, in the evening, of letting the frogs in and keeping bats out.

We talk of trees and cassowaries:

There's one plant that was growing here, we collected it soon after we arrived because we saw cassowaries eating it and ... it was a new plant and it was named after both of us Tetrastigma thorsborniorum.

No wonder they call you the matriarch of the cassowary project.

That was just coincidental that we collected it, because the cassowaries were eating it. When we came here we saw lots of cassowaries. This area had been cleared and there were a lot of guavas growing here still, they'd been spread by cassowaries, and I suppose flying foxes, and they were living all through the scrub and they'd fruit prolifically, and the cassowaries loved them and they would come each season but as the forest has grown back the guavas have disappeared.

I listen as Margaret and Kris talk stories of connection with other people working up and down the coast, stretching into the mountains behind the coast and the plains beyond. There is a man who found the body of a possum, thought to be long extinct, in an old bin at the museum. They have now located the possum in the mountain range not far from Margaret's, through knowledge travelling along the lines of the network. I see a large network of invisible lines emanating from both Margaret and Kathleen McArthur, and stretching from the coast across mountains and plains linking many people who know their bits but who are all ultimately just small knots in this net of action and meaning.

You seem to be good at networking I say.

Well that's one thing we seem to do here, we mightn't perhaps do something ourselves, but people meet here and something eventuates.

And you said you were quite influenced by Kathleen.

Oh yes, yes, when the Wildlife Society was started by Kathleen and Judith Wright and David Fleay and Brian Cluster and they started the Wildlife Magazine. *And then I think Caloundra Branch started after that and a year or so later the Gold Coast Branch started and that was where Arthur and I were involved. I always had the greatest admiration for everything Kathleen did, she is a beautiful person.*

So it was her nursery and her wildflower paintings that you knew her for?

Yeah, yeah, and then she has a beautiful speaking voice, hasn't she, to hear her reading poetry is really beautiful.

Speaking is important, having a voice in the public domain. And what did she speak—the language of the semiotic, poetry?

So after that you started growing a lot of trees?

Yes, we used to grow a lot down at Southport and we sold and distributed them for the Wildlife Society, *lots down there.*

And then we moved up here and we really have grown a lot here too, and have distributed a lot from here. We used to send big carloads up to the Townsville Environment Centre *and they'd sell them and raise funds for the centre.*

Even if you only grow one tree and take good care of it that can have quite an impact. Luckily enough the Sovereign Wood, *the big tree out the front there is one of the favourite trees of the* Torres Strait *pigeons, cassowaries eat it too, a lot of birds eat it, wallabies eat the seed when they fall. The birds are in the big tree here for weeks and weeks.*

And you've been mapping the movements of the Torres Strait pigeons for a long time?

Since 1965.

That was actually a highly acclaimed piece of research.

No just something we did, it was just pleasure (laughs).

And on the seventh day

God ended his work which he had made; and he rested on the seventh day from all his work which he had made.

Our creation achieved, we pack up and perform a ritual goodbye to the beach. At the turning circle on the headland the cars are lined up—surfies checking the surf, a couple of people taking a lunchbreak, Telecom workers managing to steal a bit of time by the sea, mothers with a car full of kids. The surf is up, there is an easterly swell as we gaze through the salt haze past wetsuit board-riders to the curved stretch of white sand beyond, until next time. We drive up the mountains through steep blue gorge country, returning to the sparse bleached-gold dryness of the tablelands.

That night, waiting for Jessie to come home, I spread all the bits of blanket over the floor—old blue and green squares, new red squares, crinkled balls of dusky pink wool and a half-unravelled vest of scarlet mohair—and begin to knit. I hear the train whistle and jump up to get Jess, scattering trails of crinkly red and pink wool behind me. Jess is amazed when she comes in, 'What are you making?' she says. It's hard to talk to an excited fifteen year old about a complex idea that I haven't really got words for myself yet. 'I am knitting myself together, making a new story out of the bits and pieces of other people, the spaces in-between.'

I show her the squares I had kept in the pillowslip, odd colours and shapes, the colours and shapes of my grandma, dead for thirty-four years, and I talk about my colours, the red colours. Jessie says I should put the blues and greens in the centre and graduate the other colours from cold to warm, 'Like a rainbow, where all the colours change and blend and merge into each other.' We talk about backing the blanket with silk dyed in lush red-pinks. She feels the soft fluffiness of the mohair on her cheek and the crinkle of unravelled wool.

I begin to knit and Jessie watches carefully. 'Can you teach me?' she says. I finish the square and give her the needles and a scrap of wool to practise with. There is something deeply and profoundly significant about this moment that stitches up holes of losses of mother, grandmother—mothering daughter: in-over-through-off—I say the words that my mother said to me and her mother said to her. Jessie follows carefully and picks it up quickly, concentrating hard as she learns to control the wool and manage the stitches, 'In-over-through-off,' she says. She wants to keep knitting, tonight we will knit words instead of reading stories. I leave her knitting in bed and the next day she tells me 'It was lovely, all the thoughts going through my mind about all the things that have happened, my friends, Dad, school, I just kept on knitting but I didn't know what I was making.' 'Squares for the blanket,' I say, as the blanket grows and changes and takes on a life of its own.

Performance VI

Houses: and the performance of home

Landmarks

I barely inhabit the dark interior of my memory of home. When I sit with my sister and imagine it now there are whole rooms that just aren't there for me. Happy inside spaces are the marginal ones, outside loo with polished lino floor where I talked to my imaginary family; a brick cubby, in far corner of the backyard; and a cliff behind where we could climb up and survey our territory. Most often I escaped to marginal places outside—the bush of Sydney suburbs in the fifties. There was ti-tree scrub with tadpole holes, stormwater drains that went under the road to the mental hospital, and groves of pussy willow so dense that old men came to live there. I knew where the best red gumtips were, when the first freesias came out and loved to pick the velvet coloured wild nasturtiums for my mother. We played in the long grass with the kids up and down the street. For as along as I can remember I was mother when I was inside. In the bush I was unbounded self, at home I was bounded mother. My work in the landscape has been a quest for belonging, searching for a sense of home in the outside world through connection with Aboriginal women's stories and then through white women involved in environmental work.

Finally, in the stillness of the cottage, I explore interior space and home, the daily inhabiting of space, and the naming of maternal space in relation to the symbolic. When I explore images of houses gathered from the other performances they

Houses: and the performance of home   181

Jessie and Natasha, cubbyhouse, 1991
(Margaret Somerville)

are houses that are differently told—letting the outside in, improper houses, cubbies, abandoned shacks, magic rainforest dwellings, witch's house. They are postcolonial houses that let the outside in; the membrane of skin of tents at Pine Gap; Nganyinytja's front doorstep in the desert; Marie's paper house held together with paste that dissolves in the rain; and a shack in the rainforest, home to a gecko, green tree frogs, and green ants. I ask the question *Is there another way in?* for women who as both home and limit/margin have no space of their own. I begin with the idea of exile from both home and language and then explore the most basic level of inhabiting—home as body, nest, shelter, cave—and inhabiting discourse. I write the undoing of mother—unstitching, unsettlement, displacement—and the daily acts of inhabiting which re-make home, leading to a reflection on Bachelard's notion of building home from the inside out: growing cooking eating and feasting food as disruption of inside/outside boundaries. Finally, I look at the gendered performance of the act of inhabiting and its relation to visibility/invisibility, seeing from the heart, and Bachelard's intimate and immense.

~

Is there another way of getting in? Below us someone has stuck four sticks in the sand and, tying the corners of a tablecloth to their points, made a shelter from the sun. This is not a primitive enclosure: it has no walls, being composed entirely of windows and doors. It does not even have a roof: the tablecloth rises and falls in the wind, like a tongue, like a wave. It sculptures the wind, crooning gently. This was the first desire, not to prohibit the elements, but to sound them, to enter into conversation with them ... It is late afternoon and the square of shadow it casts has moved a little way eastwards down the beach: to inhabit this dwelling, it is necessary to sit down outside it—just as the soul must pass out of the body and pass to the other side if there is to be conversation (Carter 1992b: 124).

Dawn

In the cottage, my consciousness unfolds just before dawn to the call of birds in trees which grow up to its walls. It begins with a slow insistent cheet, cheet, cheet, in clusters of three, then repeated; then the six-note melodious song of the pied butcher bird, the myriad twitterings of tiny wrens and finches, and the occasional screeching of sulphur-crested cockatoos high overhead.

> Listening, the analysis of hearing ... accurately symbolises what the eye tends to forget: that the body, and not only the ear, is a trembling flame, a vibrating surface, ruffled water. The body does not photograph the world, but filters it across permeable membranes (Carter 1992b: 129).

I have come to this house as an exile, another act of displacement. I need to get away from it all, I said to Laura,[1] to finish this writing. 'There's always the cottage,' she had said. It sits there for a while, *the cottage*, on the edge of my mind, while I juggle commitments and possibilities, and then I begin to pack the trunk.

What does a woman need to live, and what does she need to write? I thought of the women of *Ingelba* and their trunks:

> *Maude: Nancy's got one still now. They were just like—they never had dressing tables and things to put their things into. A trunk for the nappies and that in it. Mum used to have two or three of them. I dunno what become of them all, I know Nancy's got one there. Oh, they were nice, you could get a little key and lock 'em up. You could get a little lock and key and put them on a latch.*
> *Grace: I remember my aunty, they all had trunks, and others in Kempsey. They kept photos, papers and things their husbands or brothers made them and gifts, books (Cohen and Somerville 1990: 145).*

And Grace invited me home to see hers which had been passed through the women from her great-grandmother, Sarah Morris:

---

[1] I am especially indebted to Laura, not only for conversations, but because the cottage, and all that went with it, was crucial in bringing this work to a conclusion.

> The trunk included over sixty items all wrapped or enclosed in various containers and cloths. The objects ... included the old pipes that may have belonged to Granny Maria, wrapped in a very old hand-sewn lizard skin pouch; a collection of small glass bottles with semi-precious stones and fragments of tin from fossicking days, kept in an old handbag; Sarah's notebook with shopping list and prices from when she sent the children at the La Perouse shop in Sydney; and the letters written to Sarah from her sons who went to the war. The containers included everything from pillowslips, flourbags and old doilies, to a collection of old tins and handbags (Cohen and Somerville 1990: 146).

~

The in-between material of a life lived in the margins. Not yet literate, not entirely settled, these women carried with them the representations the new culture never offered, their writings and their photographs—and the signs of an old life now almost beyond their reach.

What can we carry with us from the past?

Salman Rushdie says of Gunther Grass

> like many migrants, like many people who have lost a city, he has found it in his luggage, packed in an old tin box. Kundera's Prague, Joyce's Dublin, Grass's Danzig: exiles, refugees, migrants, have carried many cities in their bedrolls in this century of wandering (Rushdie 1985: x).

Women, as exiles from language, may also have to carry with them the material evidence of their stories.

> This feeling of exile in relation to the general and to meaning is such that a woman is always singular, to the point where she comes to represent the singularity of the singular (Kristeva in Trinh 1991: 4).

Virginia Woolf says that to write, a woman needs *a room of one's own and independent means*. But what does a woman in exile, a migrant, a woman who is a-travelling need to take with her to write? Susan tries to reassure me that whatever I need will present itself, that is the way of writing. Maybe, I say, but I need my house journals, and they are all over the place, on scraps of paper, backs of envelopes, in diaries, sometimes on the computer, I'll never find them all. Without them I cannot write the sense of being in and of the moment; they are like the blanket, pieces of my life, I need them to stitch myself together. Yes, she says, I got to the stage where I organised and indexed all my journals; I paste in all the bits on Sundays and then at the end of the month the journal is indexed. She shows me the bulging books, containers of a wonderful collection of bits and pieces—cuttings, quotes, signs, photos, writings, drawings—and I admire this exquisite resource and her organisation. 'But invariably,' she says, 'when it comes to the actual writing, it's what's in my head, in my body, that comes out.'

Trinh says:

> that an exile can be worked through as a crossing of boundaries ... Writers who in writing open to research the space of language, rather than reduce language to a mere instrument in the service of reason or feeling, are bound like migrants to wander from country to country ... they disturb the classical economy of language and representation and can never be content with any stability of presence (Trinh 1993: 7).

So I pack *The Moosewood Restaurant Kitchen Garden Cookbook* for Laura, and some seeds, a computer and printer, my daughter, and some clothes and drive to Laura's cottage in the Warrumbungles. The next day Laura and I sit in the garden under the trees and eat lunch of brown rice and salad. We talk about what one needs; of bodies and journals, clean food, clean water, growing vegetables, cooking, gardens and compost. Well, I say, if I don't find my journals on the computer I'll take my camera and spend a few days in the sandstone caves.

What is the most basic level of inhabiting? Bachelard says that 'all really inhabited space bears the essence of the notion of home' (Bachelard 1969: 5).

Kris: The inside and the outside. We have an outside covered with skin, skin with a number of layers, the most outside layer continually sloughs off cells which are replaced from underneath. Our outside wears away and as it wears away it continually grows. The skin is like the walls of a city or a heavily fortified house, but there always has to be a way in and out for the outside provides the substances for maintenance of form and renewal. Under the outer layer are nerves and blood vessels, an array of sensory cells in muscle and under the skin in the tissue which ties everything together. And of course hair. Hair reaches out from the inside to the outside—it protects us, it reflects our state of being to others outside us. Epithelial tissue, which lines all the boundaries and tubes, is made up of cells which selectively allow certain matter to pass through the cell into the blood vessels and around the body. Matter and information pass through the cells and the cells themselves have surfaces for interaction. These surfaces are at the margins, on the outside of the inside or the inside of the outside, and it seems that all activity, all generation and regeneration, all exchange of information and matter take place on these surfaces (Plowman 1994: pers. comm.).

In the desert there was a *wiltja* of intertwined mulga branches at our head, a fire of mulga wood at our feet and the shelter of mulga trees overhead. The word for camp (home), is *ngurra*, the same as the word for country.

> People coming to a new location could humanise it as a camp within a few minutes; a windbreak set up, fires lit, gear stowed and food cooked. Such a camp (*ngurra*) made by individual people never loses its original association with them ... On the other hand ngurra as country *is beyond the individual human experience, it is considered to be immutable* (Jones 1991: 31) (my italics).

Near the Warrumbungles, in the Pilliga Scrub, the sandstone caves form a stone knoll on top of a hill. Their outside skin is weathered to an old and wrinkled grey, covered with lichen and inhabited with grasses, trees and bushes growing from its surface. Throughout its kilometre or so diameter, the knoll is pitted and pocked, honeycombed and gouged through with tunnels, overhangs and shelters of every size and shape. Inside, its skin is soft and new, folds and curves of flesh coloured hollows, delicate pink to pale ochre. In places, the effect of weathering on the soft surfaces is so new that small handfuls of coarse pink grains lie at the base of the hollow, and in larger caves there is a deep bed of fine sand. In the day, all appears empty and silent, but on closer inspection every groove and hollow reveals traces of inhabitation. There are paw marks, claw marks, marks of sliding and hopping, and the tiniest of tiny bird prints; nests, webs and beds shaped into sand and tunnel; bodily extrusions and dung of every variety. It is a place of play; of climbing, crawling, sliding, jumping, hide and seek, of peering through holes to the world outside; and of fantasy. It is a journey through the body's interior surfaces, down the trachea or oesophagus, around the curls of the ovaries, floating in the uterus or stomach, bouncing in the cushioning of the lungs.

The sandstone caves are closed to visitors from Spring to mid-summer for the nesting time of the rare Peregrine Falcon.

I visited the 'archaeology' caves in Coonabarabran to explore the feeling of inhabiting these spaces. My first impression of Kawambarai Cave was one of 'felicitous' inhabitation, cool in summer and warm in winter, protected from rain and wind, but now deserted. There is a deep and eerie silence; even the wind doesn't enter the cave to disturb the stillness.

~

I sleep in the cave on another occasion in mid-winter with son and friend when the track is too wet and we have to leave the tent behind. This time the creek is swollen and the weather freezing. To reach the cave we have to take off socks and boots, roll up trousers and walk in icy water. In the black night we are careful not to overbalance on shifting rocks in fast current. In the cave the little light of our damp and scratched together fire is welcome and we fill hot water bottles with a cup of water from the billy and make a body hollow in the dry dusty floor to sleep in. I feel deeply safe, womb-like, as if, being part of the core of the earth, I am beyond harm. But there is a sense further back in the cave of danger, of being sucked into the depths of the earth. We share our space with a bat who flutters almost imperceptibly above our heads and my night is peopled with many strange dreams. One, in which I am lying with a baby in the curve of my body, and when I wake the baby is dead.

~

I asked this question, What is the most basic level of inhabiting? of the women in *The Sun Dancin'*. We went to various caves together, mainly in association with archaeological digs, and the question became one not only of inhabiting caves but of how they inhabit the various discourses, such as archaeology, that provide 'othering' representations of themselves.

Madge: I got a bit frightened first because of all the people, not that I'm frightened, but when you meet strange people you know I was a bit shy at first but then after I got to know the people talkin'—and they was talkin' the same language, talkin' about Aboriginal people. And when they talk about them they just ask about this and ask about that. I didn't feel shy any more because I knew that they were just there lookin' for the 'istory that we wanted too, you know about our ancestors and it just felt like my own people because they were there to find out all these things. And I wasn't a bit shy. After you left I got in there and I was diggin' and I weighed all the dirt then shook it all out and took it down and washed it really good. Exactly like gold minin' like you see on TV, olden days wasn't it—you sieve it and then you put the water in it, so—it was good.

Rather than become embattled or territorial towards the places or talk of archaeology, Madge finds a point of intersection with her own story and moves in to gently inhabit the archaeological space.

Madge: It was lovely, I was just sittin' thinkin' about all the old black people you know and the kids playin' in the dirt and old people, probably the men makin' things, I could picture that. It was lovely, a lovely feelin' and I don't like dark things like that but it was nice wasn't it. I said to young Mick I could go out there and live in the caves you know it was lovely, quiet, not any bit of wind, beautiful.

Madge's 'babies' are still alive, and she uses fantasy and imagination to enable her to inhabit the space and language of archaeology. Rather than a gesture of colonisation (I know better than archaeology) or a retreat to being colonised (archaeology is the expert), Madge teaches me a different relationship to the discourse and discursive practices of archaeology.

~

Morning rising

At the cottage in the morning I pick up sticks from bird trees and light the pot belly because it's still cold in the mountains. The fire crackles as the house wakens, and warms my back while I do morning yoga. The kettle boils and I make tea for my sleeping child before I go out to walk.

I wonder, as I stretch, what is central to this act of inhabiting, and imagine the morning routine Patsy describes in *Ingelba* centred around her grandmother's hearth.

Then you'd look over one end and you'd see this beautiful big hearth made from stone—stone, mud and clay. It was one length of the wall, one whole wall was fireplace and the stove would be at this end, the rest'd be just one big hearth, whitewashed every day. The whitewash came from the spring just down from the laundry, some really white clay there. That was the hearth and the logs used to be from one end of it to the other. Pop would have big back logs like that and one big back log would go all night. We'd get up in the morning about half past six and get ready to go to school and they'd still be burning and the old black kettle'd be singing away.

Patsy describes the building of hearths as if they were the building of the whole house. It is the only building activity she talks about, even though she describes all the houses as handmade.

The hearth was about four or five feet high and about eighteen inches thick. I used to watch m'grandmother making them. After a while, especially with the wood that Pop used to get, it'd be box and stringy bark, after a while it would burn out, the hearth would start burning out like iron dust. She'd patch it up then, every so often, patch it back up with the clay and the mud to fix it all up again and I watched 'er doin' it.

Pop'd carry all the rocks and stones 'cause she'd have big rocks and stones in hers, I only had little ones that I could carry. Between her and Pop they'd put them in place and mix up the clay and just pat it on the rock like that and make it all smooth like building a wall and filling the gaps in. Filling all the gaps in and then whitewashing it all over, which'd keep it together as well. Everyone sat around the fires and it was a cold place in winter, very cold. You'd get up to −14 there in wintertime 'cause the river freezes, you get ice on top of the river. But I don't really think of the cold part of it 'cause it was a warm place for me.

The hearth is the symbol of her grandmother's love and the centre of warmth for Patsy. Not only was the inside of the house organised around the central feature of the hearth, but at Ingelba the position of houses was organised spatially around the configuration of the (grand)mother's hearth. The little shacks of aunties and cousins radiated in circles around the grandmother's house, bounded by the curved line of the river. The hearth was the most substantial feature of the house, and is the only mark that remains on the landscape now that the people have gone. They represent the patterns of life through which the people can read their story.

Gaston Bachelard describes the house itself as being synonymous with the maternal.

> In the life of a man [sic], the house thrusts aside contingencies, its councils of continuity are unceasing. Without it, man would be a dispersed being. It maintains him through the storms of the heavens and through those of life. It is body and soul. It is the human being's first world. Before he is 'cast into the world', as claimed by certain hasty metaphysics, man is laid in the cradle of the house ... Life begins well, it begins enclosed, protected, all warm in the bosom of the house (Bachelard 1969: 7).

House as womb, house as body of mother. In Patsy's description of hearth-as-mother and mother-as-hearth, the mother is actually seen as *constructing* the hearth as well as *being* the hearth. The literature quoted in Bachelard, and much of Bachelard's own writing, leaves woman stranded mute, identified with the house itself, just as Walker describes Irigaray's idea of woman entombed at the centre of philosophy.

> The maternal body becomes metaphorically the silent place of philosophy, the backdrop of all representation, as the silent unacknowledged site of the philosopher's being ... the mute stage or theatre of his representation (Walker 1993: 5).

Barbara Holloway, in her thesis on the inscription of land in the new colony, consciously sets out to unravel the knot of woman as both home and envelope/limit, in her analysis of late nineteenth and early twentieth century Australian poetry. She argues that when faced with a new and foreign landscape, it was the trope of woman that served to make the landscape familiar, storyable. She shows that in the inscription of space into place, so necessary for the establishment of both material action and textual representation in the new land, *woman* provided both inner core (as home) and outer limit (as frontier). She refers to the work of Bachelard to theorise woman-as-home, to Kristeva (and the abject) for woman-as-limit and to Burgin's view of the pre-oedipal mother as the first object of abjection:

Significantly, the first object of abjection is the pre-oedipal mother—prefiguring that positioning of the woman in society which Kristeva locates, in the patriarchal scheme, as perpetually at the boundary, the borderline, the edge, the outer limit, the place where order shades into chaos, light into darkness (Holloway 1993: 15).

Holloway (1993: 24) names the difficult and challenging question: *How, then, are women to relate to place if they are the maternal and if they are also the envelope/limit?*

Unstitching

There are three small images that I carry with me. The first is a miniature tapestry of an English country cottage and garden. In finely stitched thread there is a thatched-roof cottage surrounded by an abundant cottage garden of hollyhocks and foxgloves, big soft purple irises, columbines and pansies. The woman stands on a winding stone path (there are no weeds in this path) with large sun hat and long skirts trailing into flowers. The colours of this tapestry are the soft pastel hues of spring. Viewed closely this picture is composed of a million tiny woollen threads which are woven together intricately to produce the seamless effect of the whole.

> If traditionally, in the role of mother, woman represents a sense of *place* for man, such a limit means that she becomes a *thing*, undergoing certain optional changes from one historical period to another. She finds herself defined as a thing. Moreover, the mother woman is also used as a kind of envelope by man in order to help him set limits to things (Irigaray cited by Holloway 1993: 114).

In the second picture, the woman has moved down the garden path towards the hills beyond. She wants to step out of the tapestry, out of the beautiful garden, but each of the tapestry threads, in their delicate pastel hues, is stitched through her body. When she moves down the path towards the hills, the cottage with the thatched roof, the garden path, the hollyhocks and foxgloves, the soft purple irises, and the columbines, are each dismantled, as, one by one, the woollen threads are pulled through her flesh.

> Although Foucault's notion of 'inscription' implies writing on a surface, this surface is, as in Derrida's notion, an exceptional one that includes the body's depths and, indeed, questions the division between interior and exterior. For example, Foucault argues that genealogy traces the workings of history as it 'inscribes itself in the nervous system, in the temperament, in the digestive apparatus' (Kirby 1991b: 21).

In the third picture, the woman stands out of the frame, naked. Beside her in the frame are the fragments of her memories, a few photographs, some old patterned china, bits of her mother's jewellery, some embroidered cloths. Her body is full of holes and where each of the tapestry threads was pulled from her skin, stitch by stitch, there are marks of blood on her body. Of the original picture, only the pansies remain, pressed flowers at her feet, the colour of blood, and of passion.

Unsettlement

> Dear L
> A fast red Falcon and a tank full of petrol is just about all a girl could want, a few knickers and endless miles of coastline, on one side blue sea with blue-green islands rising from it, and on the other blue-green mountains rising from the hinterland against the western sky. At Mission Beach it was the fourth day after the full moon in spring, the day of coral spawning. The male and female gametes are released from the coral into the sea where they float on currents and tides in their millions. They form trails of gold glitter as they float on the surface of the sea in swirling bands. I swim in these primeval waters and wonder how my body would respond to the moon if I lived on a beach by the sea.
> Love M

We stop at Harry's shack at Trinity Beach. A tiny timber house almost hidden by huge trees and a massive stand of bamboo. It is tinderbox dry and the leaves crunch underfoot inside and out. Inside is bare except for two wire bed frames, an ancient cooker, sink, bath and toilet, and leaves everywhere; several of the walls are made of netting. I am grateful for the small movement of air in the hot still evening as I colonise the space by sweeping leaves and cleaning the loo, overrun by green ants and green tree frogs. At night, wind rustles in dry leaves, insects flicker around the light and persistent crickets sing to the night in a wooden house that is as light on the land as the trees themselves. A home for a gecko, a skink, abundant green tree frogs, millions of green ants and it seems, a host of unseen termites who will return it to earth as quickly as it appeared on this landscape. Just a change of form.

Moving through the landscape in North Queensland, moving from place to place, different people and different places each night, no returns, I ask what is *home*, and what is this process of unsettlement? Of all the millions of new coral zygotes, only a few will find the right conditions, settle and grow into mature coral, the rest will perish and die. Is this settlement necessary for all of us to grow and flourish?

Displacement

I first read Bachelard when I was in the Women's Shelter with a strange mixture of nostalgia and anger. Nostalgia for the home I had left behind and anger at the unrecognised work of woman in constructing that home. The production of building the house is always obvious; the production that turns that space into a home is invisible. And 'home' for many is not always a 'felicitous' space, it is not necessarily even safe. I wrote in response to Bachelard:

Over the past four weeks I have fled my new 'home' twice. Like a refugee, hurrying to escape immanent danger, hastily packing our things for fear of being trapped in my own space. I have never had time to settle in my new space, to send down a little root to nourish and protect. I learn a process of displacement. This place smells of leftover milky weetbix, nappies and stale cigarettes; fear and despair soaked into the long shag pile of fine Cape Cod architecture. Rosa calls it the refugee house and says she feels like she is in prison. There are bars on the windows and the doors are always locked and I learn to carry keys everywhere for the first time. I remember keys when I worked in a remand home for juveniles. Am I in gaol? But then I lock myself into my room with daughter and dog, two mice and a few belongings and I think maybe I have with me everything I want.

I decide to interrogate Bachelard from the position of the dispossessed. The image that continues to haunt me is that of the *old faithful servant* quoted from the writings of Henri Bosco:

> This vocation for happiness, so far from prejudicing her practical life, nurtured its action. When she washed a sheet or a tablecloth, when she polished a brass candlestick, little movements of joy mounted from the depths of her heart, enlivening her household tasks. She did not wait to finish these tasks before withdrawing into herself, where she could contemplate to her heart's content the supernatural images that dwelt there. Indeed, figures from this land appeared to her familiarly, however commonplace the work she was doing, and without in the least seeming to dream, she washed, dusted, and swept in the company of angels (Henri Bosco cited by Bachelard 1969: 68).

Where, I ask, is *home* for the servant? Is she happy existing on the end of someone else's fantasies or are the dreams an escape into her own? Who owns the house, who tells the story and whose interests does the story (house) serve?

Morning walk

Each morning I leave the cottage and breathe the air of trees. Even in the dryness of drought, their moist breath enters mine. I walk the well-worn path to Laura's house, crunching of shoes on gravel, then down the hill and over the creek. We walk around the paddock, three times, repeating the marks of our patterns of movement through grass. Every morning it is the same and every morning it is different.

This morning the rain clouds are gathered in the west around the mountain (Malcolm's mountain, 'up the mountain' he says, which mountain is this?) and fine misty rain barely reaches the ground. Today the palest pink buds on the quinces are further out and the air eddies differently with this light misty rain and the cold of no sun. We talk about Bachelard's 'function of inhabiting'. It is this function of inhabiting that throws into high relief the process of this writing. I have to return again and again to the performance of inhabiting, to follow it in my body, to know it in and of the moment.

I talk to Laura, through the notion of performance, how one inhabits space as a performance, the dailiness of it, and we talk of Virginia Woolf and her feeling like 'a fish in a stream, deflected, held in place but unable to describe the stream' (Woolf 1976). Is the stream, I ask, the mother? And is this why she cannot describe the stream, because her mother formed the fabric of her life? Is this action of inhabiting, the production of the fabric? And this relates back to my original question, How can woman-as-mother inhabit the landscape when she is the landscape? Where is home for mother when mother leaves home?

After the morning walk there is a flurry of busyness. I make breakfast of grated apple muesli. Place a handful of rolled oats into each bowl; peel and core two apples and grate over the oats; squeeze two oranges and pour the juice over the apples and oats; add a handful of chopped almonds and a tablespoon of yoghurt. Wash the dishes, sweep the floor, wash some clothes and admire the washing on the line lit by the sun, hanging between white-trunked trees. So many women have commented on the pleasure of a line full of clothes. 'Where does this pleasure come from?' I ask Laura.

We talk about the pleasures of the washing line—'Sometimes,' Laura says, 'I hang the washing out in segments and sometimes I go right around the outside of the line. Sometimes I start from the outside and work my way in and sometimes I start from the inside and work my way out.'

And we talk about how our mothers used to hang out the washing in meticulous order and Laura says 'Yes I like to do that too; sometimes the classification is by people and sometimes it's by items of clothes—all the shirts go together, the socks. It's a bit like a performance,' she says, and tells me about an artist who performed the washing. She calculated how much washing she did in 365 days and she did it all as a performance, washed, dried, ironed and folded. First she did it all in white and then she did it all in blue.

I remember Maude in *Ingelba* talking about the sweet smell of sun in clean washing:

They just had coppers and they'd boil the clothes up. They'd be lovely, especially the towels and sheets and pillowslips and things, when you bring 'em in after boiling 'em all up, lovely fresh smell, when you get inter bed with 'em.

And Marie, in *The Sun Dancin'*, talking about their daily household routine:

Monday Tuesday Wednesday Thursday Friday we had to work, scrub the tables with sandsoap, get the soap and put it in the copper and boil the sheets and put 'em out on the line. If they had one mark on 'em we had to get 'em off and take 'em back. And each clothes had to be pegged by one peg only, every sock, and Mummy come along and check it, and if it wasn't right and if we didn't do it we'd get the wattle bush, we used to get a bashin' eh mate. We'd make the brooms eh. What about the day we brought the ti-tree home and it was young (much laughing). 'Don't you know what to pick?' Mummy said. I said, 'Yeah, I knew what had to be picked.' We'd roll it up with the twine then you had ter go and chop a little tree for a 'andle. We 'ad to do the work, so we brought home the young stuff and all the flowers was on it and when we swept the house out it was all flowers and petals (laughs).

Everything was well kept 'though it was a paper 'ouse.

And Emily wanting to make the graves visible by sweeping the earth around them.

Is this the act of inhabitation, the intimate tasks that we do each day? Is this what Emily was doing when she performed the failure of visibility, *the graves have not been swept clean*, was she re/marking these acts of inhabitation which made women visible in the landscape?

Bachelard writes about women's work in the house as building the house from the inside out:

> A house that shines with the care it receives appears to have been rebuilt from the inside; it is as though it were new inside. In the intimate harmony of walls and furniture, it may be said that we become conscious of a house that is built by women, since men only know how to build a house from the outside (Bachelard 1969: 68).

We talk about turning the inside out, like the washing hanging so publicly at Pine Gap, the making visible of self-in-landscape. We talk about making houses from the inside out and embodiment and a community of washing and the excesses of curtains, calico on the line. 'It's the potential for production,' Laura says and I say 'Isn't that the same as the washing, it's about cloth, the folding of cloth, the dressing of a body and the dressing of a house in curtains.'

How does one make a house from the inside out?

When I buy my house I am still full of holes.

There is a huge gaping hole in the back wall where there will be French doors; the side walls have been ripped apart and ends of plumbing fittings and electrical wires hang out; next the entire kitchen and living room floor is jackhammered to rubble; I wonder whether it will ever come together again.

A body needs a shell around it, an outer skin extruded from its membranes, like the snail's, to protect its softness and vulnerability from the world. My friends come and seven of us scrub, fill and paint for two days, making soft flesh coloured walls, the colour of new baby's skin. I make skin over bone of concrete floor, soft tawny pink cork. Eighteen packs of cork tiles, a straight edge, a carpenter's pencil, an adhesive spreader, two tins of adhesive and three of polyurethane. I begin with a cross that extends from side to side the length and breadth of the room to run the rows from; fill in each of the segments of the cross; fill the spaces in between and measure and cut the edges. Three coats of polyurethane for protection.

I find some fine unbleached hand-loomed muslin, buy three metres, and clothe my house, beginning with bloomers, muslin half curtains on the kitchen window. Creamy off-white and fine-textured, like spider's web across window glass; frills top and bottom and a soft one-and-a-half-times gathered, my first sense of enclosure, a soft inside space.

I have a new meaning for Trinh's 'get dressed and stay at home'.

Then my whole house is draped with long swathes of papery unbleached calico as I sew and stitch together the day's events. With each curtain I consider the folds and length and feel of the room. The colour is soft, unbleached and they have a tempting papery crispness, empty canvases, blank sheets of paper waiting to be inscribed. I write on some of them, from Body/Landscape Journal the Last, writing to stitch up a life. Each curtain adds to my sense of inside space—there are now curtains in bathroom and bedroom as well as kitchen. I think about the layers of insides and the desire to make a house for myself out of fabric, perhaps starting with knitting because then I can spin the wool too. A large tubular curly extrusion of reds, oranges and yellows like the curls of body tubes, moulded to body shapes.

> Dreams of a garment-house are not unfamiliar to those who indulge in the imaginary exercise of the function of inhabiting. And if we were to work at our dwelling places the way Michelet dreams of his nest ... each one of us would have a personal house of his own, a nest for his [sic] body, padded to his measure (Bachelard 1969: 101)

Mid-day

In the cottage Laura and I share lunches and talk. Mostly I make a plain lunch, brown rice or millet, with pumpkin, sunflower and sesame seed (ground), and fresh salad. I search out the best quality organic or biodynamic vegetables to buy, or grow them. Tear the leaves of fresh green lettuce, grate crisp carrots, cut thin slices of red capsicum and finely chop a bunch of parsley. Toss, and add a squeeze of lemon, dash of oil, sprinkle of cumin. We eat inside or out, depending on the weather. Sometimes at the weekend we go for a picnic and sometimes Laura makes a feast.

When my babies are little I breastfeed them all. I eat, digest, and process my food, reproduce it as milk and offer it from my body. For a long time they eat nothing but breast milk and I love the tingling pleasure of their sucking. Their baby shit smells of warm malt, slightly sour when they are distressed. As they grow older I plant seeds in moist soil, feed them with compost and water, harvest and prepare them for eating, returning the waste to the soil.

But thinking about food now there is a dream. In this dream I want to finish my writing but there are many visitors and many mouths to feed. I am writing a message list for a woman who has adopted delicious twin babies. They have just started eating solids—so I write *apples, bananas, pears*—and I suggest, *milk arrowroots*. But I write the list in fine black pen on a slice of multi grain bread and then I eat the slice of bread.

> It was through learning to cook that I finally broke into the world, joined it and was joined by it. At first food was just a drug to dull memory and pain. Then I discovered that flour and butter working between my fingers, carrots falling apart under my quick knife, egg whites rising in the bowl as I beat them, all began to give me a sense of my adequate power, my reality. Inventing then writing down recipes I unmade and remade the world, and, after a childhood in which I could not trust others' words, learned to discriminate, to speak. I discovered that it was important to know how to select and classify, just as it was important to know how to combine, to unite, to merge. Through cooking I learned how to know when to separate and when to merge. Through cooking I learned how much I was part of the world around me, no different from leeks and peas, and how much I was separate from it, person not leek (Roberts 1991: 87).

～

Often after lunch we go to check the tiny seedlings of carrots, lettuce, radish, coriander, coming up through brown earth little by little each day. We water newly transplanted lettuce, tomatoes and basil. At weekends we collect manure and make compost, dig new garden beds and make garden plans. Sometimes we just sit in the garden surrounded by sparse dry paddock ending in the dark casuarina curve of the creek and dream garden dreams. Garden dreams that inspire cassowary women to look beyond their garden fence and grow rainforests as well as food.

Every day I love this garden but it is, I often think, a mark of colonisation. It uses up water from fast diminishing pools where water rats and turtles live; the food that grows there is imported with our European origins. If we really knew how to tread lightly on this land we would eat kangaroo, and goomeyii, five corners and fiery hogs, the eggs of the turtle and perhaps echidna and possum.

The women from *Ingelba* and *The Sun Dancin'* told me how to forage and cook these things.

～

Laurel: I was tellin' you about the possum—my father used to do 'em lovely. He'd skin 'em, they're very hard to skin, he'd cut 'em all up in pieces like you have pieces of chicken 'cause their bones are smaller than a chicken, and he'd put them in a pot, put a lid on it, put curry in it, a certain amount of water, then he'd wait till that got really nice and tender, chop 'is onion and potato up, and by the time the potato went through—the potato used to thicken it a little bit—it would melt in your mouth. He was a beautiful cook.

Violet: Yeah, grubs are the best meat out. You ever eat fish? Well they're nice cooked. Those big gum trees, we used to go out to the gum trees, we'd see all the mess around, we knew there was a grub in there, we get a little axe to cut it out, get a piece of wire and pull the grub up out of it. And when they're grilled, take the inside out and eat 'em with pepper and salt, they're better than fish.

Maisie: They used to get the red fish and we used to always get those mussels too, they were like the pippies from the beach. Cook those on the coals too they used to. Turtle's egg too, they used to eat a lot of turtle's egg. When we come home from school, go to the riverbank there, the crick, and dig the anthill where you always get the turtle's egg. Boil 'em or cook 'em in the ashes. You ever cooked egg in the ashes?

Mavis: We used to eat fiery hogs, five corners, geebungs, slipstones, all that and the gum off the wattle. I used to love the gum off the wattle because we'd wash the gum and put it into a cup and boil it up with water and it'd come out like jelly.

~

The cottage is surrounded by bushes of fiery hog and five corners and today the yellow orchid flowers of the goomeyii appear which means it's time to harvest the two white bulbs underneath. The garden around the cottage is just the barest mark, a winding stone path, and some low loose stone walls that follow its shape. At the foot of the large silvery white trunks, some scented geranium grow lank and tall to reach the sun, a straggly bush of lemon thyme and a few old roses. I like these hardy old survivors. I make a corner for the compost bin and plan a little planting, just a purple flowered sarsaparilla, that grows here anyway, some hardy bushes of rosemary and lavender—or perhaps I will just watch the bush take over and grow sprouts in a jar on the windowsill. From the window I see the cleared patch beside the house where some trees were taken out for electricity lines, the scrubby undergrowth has already colonised the open space —five corners and fiery hogs are thriving in this new patch of sunlight.

One day in the cottage garden Laura sets tables with fine white cloths in the dappled light of white trunked trees. There are salads, pink salmon quiche, truffle oil beef, trifle, hazelnut and strawberry cake and cool drink. It is our feast day.

I ask Laura about her truffle oil steak.

You take a whole piece of meat, trim off every skerrick of fat and sinew so there is no barrier to the melt in the mouth. I usually buy best eye fillet and make it nice and round, tie it up in a neat parcel with best linen thread, offcuts from the loom. Then marinate overnight in truffle oil, garlic, sea salt and balsamic vinegar. Truffle oil smells and tastes unlike anything else, so of the earth, rich deep and brown and the balsamic vinegar is the acid equivalent of truffle oil. It gets laid down as dowries, fifty year old, ten year old—and the bottles are part of it too. It is the whole performance, I usually keep the bottles. Once the truffle oil and balsamic vinegar go together, they are both incredibly rich and complex in their own right, everyone in the house comes and has a sniff. You cook it very fast in very hot oven, it should be pink when served, and at room temperature— on big china platters with silver servers.

~

In the Bunya Mountains when the huge fruit of the bunya pine ripened every third year, there were massive feasts and ceremonies when people gathered from far and wide. There are stories from the Darling River of huge hunting nets which would take months to make and were made in a large-scale joint effort by both men and women. The amount of labour required for this production was so great that they are thought to have been only used for celebratory occasions when large numbers of people had to be fed. In the central desert big ceremonies could only be performed at the time when the grass seed was ripe and there could be enough grain. The economy of the ceremonies involved a complicated exchange between the women who collected and prepared the grain and owned the grinding implements and stories, and the men who wanted to perform the ceremonies.

At Ingelba and Burrabeedee big feasts were prepared at Christmas time.

Emily: At Christmas time we'd all meet at Granny Robinson's place. There was Mum, her daughter and the daughter-in-law. They'd all meet there and they'd all set jellies. There was no ice chest, there was no refrigeration. They used to go into town to the freezer works and get big tubs of ice and they'd set the jellies the night before in the drip safe with the water running down the bag and they used to have big hot dinners and the camp ovens was that big you could put a full pig in. Just cut 'is back to make him round like that and stuff 'im up and put him in one oven and a full sheep in the other one and they'd have these—you know the tins that they used to call 'em kerosene tins, they'd buy 'em months before Christmas, scrub 'em and they'd be shiny and clean, handles on 'em and a big thing right across the fireplace like that. There was a bucket of peas, there was a bucket of beans, there was a bucket of cabbage and they used to put the big hot dinners on and then the jellies and everything set perfect. So that was a big day. And that table would be set at Granny Robinson's place for one week. Anyone that come to Burrabeedee to the Mission there was a place at the table for them.

~

Gay Bilson, whose performances highlight the body (of woman) at the site of food preparation, writes of the time she planned to make blood sausages from her own blood for a dinner party, 'the ultimate symbolic gesture of generosity':

In 1992 I approached passing pathologists and lawyers, wanting to make sausages from my own blood for a dinner centred around the body, our bodies ... Our blood has similar properties to pig's blood so as 'food' I knew it would be palatable. Poaching them would seal the safety factor. Offering them (the consumers would have full knowledge of the blood's provenance and so be able to refuse) would be the ultimate symbolic gesture of generosity. Having put the idea to a crucial group of cooks and servers (even though in one sense I was not inviting them) I found gut resistance on the most part which seemed, at the time, insurmountable (Bilson 1994: 68).

Is it, she asks, because the gastronomic arts are so truly participatory that there are such strong taboos? Or is it, I ask, because this exaggerated performance of the body of woman as the ground on which we all depend is not palatable because it crosses too many boundaries?

Midafternoon

The cottage seems to be composed only of windows. I live in the light and space of these openings. In the early morning the sunlight casts a pattern of fine needled cypress on the wall where I work; then as the sun moves, leaves of eucalypts dance around me in dappled light; the afternoon light refracts off white-trunked gums from the back of the house; and at night, in the light of the moon, the scribbly gums shine silver. *I see only through the touching of the light* (Irigaray cited by Grosz 1994: 106).

I came to the cottage to escape the eye. It was the eyes that bothered me, the eyes of other houses too close, looking into mine. The eyes at the office, expecting, asking. The house began to feel like a prison, the dull sounds of suburbia and the isolation of closed walls that represent neither connection to outside world of bush or sea, nor to other people. The all-seeing eyes of all expectation but no connection.

In the cottage I gulp the air of solitude in great hungry mouthfuls and plant vegetables for the time I can avoid town. Every day is so routine that the vagaries of weather, sun, moon and stars, the movement of ants on the path by the door, are noted, and I go deeper into retreat than I thought possible. In the stillness I read *The Orchard*:

> It was as if, in the dimming of my sight, I realised for the first time the exhaustive, exhausting monitoring that is made of us wherever we go. Eyes that look, that see, that watch ... I looked at myself in the mirror that day and saw that the inability to see how others saw me ... was a dark blessing: for a moment I had stepped, or been stepped out of the onerousness of that exchange (Modjeska 1994: 117).

Modjeska writes about her realisation of the extent to which she was required to perform herself for others when, in losing her sight, she could no longer see herself as others saw her. She could no longer participate in the normal economy of social exchange. In her enforced retreat she says *what is required of us at times is not performance—that endless dance of display—but the simple task of being* (Modjeska 1994: 118–19) (my italics).

For me this retreat to the cottage is one of choice but it is also about escape from the burden of visibility and expectation; it is the refusal of performance for others. There have been two retreats since I began this writing, the first in illness, in which there were no words, when, like Modjeska, I simply learned to be. The second, in the stillness of this present refusal, I begin to reflect on the daily acts of inhabiting, to name, first of all for myself, the space of intimacy. For how can I reflect on the performance of self for self when the performance is always for others? In this retreat a way of seeing can be performed that makes visible the invisible ground of the function of inhabiting.

The paradox of this performance through retreat does not escape me. This paradox of the making visible the invisible was experienced most forcefully on the Mission Beach trip when I recorded in my journal an overflow of excitement and ideas about what I was seeing in the landscape. But each time I went to a meeting of the Wildlife Preservation Society of Queensland, ostensibly about conservation, where the predominant discourses were of science (of the species), my images and words disappeared. For days my words were no longer accessible even to myself, a process I described as invisibilising. It was as if, when confronted by an objectifying discourse there was a choice of two positions, the objectifier or the object and I didn't want to accept either. Later, in solitude, the subject responses of my embeddedness in this world of landscape and people would re-emerge, a way of representing that didn't involve separation of seer from seen.

The same paradox of performance is apparent in the constitution of the cassowary project. The project relies on the visibility of the normally invisible bird. In fact the bird must retain its invisibility to survive, but in order for the public to accept the need for protection it must be made visible. The methods for the renewal of cassowary habitat rely on the invisible (abject), a movement from the miniature and unacceptable of a cassowary dropping to 'the immense' of a rainforest. Similarly the act of inhabiting—one's body, or one's home—necessarily remains the invisible ground on which all thought and action is based.

This paradox of performance is not something I can think through by conventional means. I had a letter from Paul Carter in response to the Emily chapter: 'My sense on the basis of a quick read is that you have yet to "perform" your stories in quite the way you desire. But as the scenario of a performance *Emily* already works very interestingly.' I thought carefully about what I wanted of the concept of performance. Should I, I wondered, remove the word performance from the chapter headings? Perhaps they are explorations of performance rather than performances themselves. All I could say in reply was that performance was different for men and for women. And then I awoke one Saturday morning with the idea that the question is one of *the gendered nature of performance and the performance of gender*. In Carter's scenario, am I being relegated to the fabric again? What if the 'scenario', the making of the fabric, was the actual performance? For women to make visible in the landscape, or in their stories, what would normally be invisible, is already a performance. It seems to relate to the general invisiblity of bodies, and for women to the 'seeing eye', the 'I' and the 'i'.

> A man writes 'I' as he sees, and in writing it is therefore seen. The relationship is clear. When a woman writes 'I' she must reconcile seeing with being seen, and negotiate the transposition of the first term to her own use. How is she, the object who is seen, to see herself, both seen and seeing? (Modjeska 1994: 141).

Modjeska develops this idea further when she describes what she calls *seeing from the heart*. She remarks on the irony of losing her sight just as she begins a project on women's art. Unable any longer to distinguish the fine detail of the paintings, she has to learn to see in another way and it is this other way that she describes as seeing from the heart. To me, it is a seeing without separation, without the distancing of seer from seen. One of these artists whose work she reviews is Grace Cossington Smith whose interiors she describes as:

> self portraits of a woman who has resolved the tension between her own ability to see and the seeing, or being seen, that is required of her: a woman who has fully withdrawn from the gaze of the world to discover not a defensive retreat, but the fullness of a solitude that society deems empty (Modjeska 1994: 136-7).

Modjeska charts the development in Cossington's Smith's work from a grim self-portrait, in which 'seeing and being seen are held in painful tension, a dark and punishing solitude that contains as much refusal as release', to her interiors-as-self-portraits. She describes one of these, 'Interior with Wardrobe Mirrors':

> In it the mirror of a wardrobe door swings open in the centre of the painting, where it invites our own reflection—and in that invitation we see the absence of the painter whose image should face directly into that shining surface. Instead it reflects a door which opens across a verandah, across a lawn, to trees and a distant sky. Where the artist should stand, stands instead an invitation to the world, to all that is beyond. That is the fullness her solitude has produced (Modjeska 1994: 137).

In this I understand Trinh's 'expansion through retreat', a movement inwards generates a movement outwards towards the world beyond:

> Edenshaw's statement remains multi level, it ultimately opens the door to a notion of self and home that invites the outside in, implies expansion through retreat, and is no more a movement inward than a movement outward towards others (Trinh 1993: 5).

The view through the open door to the world beyond called to mind many houses from this long story. Are they also many self-portraits, many versions of 'an invitation to the world, to all that is beyond'?

There are the tents at Pine Gap with no walls and a cloth ceiling that colours the light; the camp Nganyinytja made for us in the desert, *wiltja* at head, fire at foot and branches overhead; Patsy's grandmother's house where she can see the stars through holes in a tin roof; May's house where doors might be in a different place when the family comes home, and Marie's paper house with walls held together by flour and water that dissolves in the rain; Margaret Thorsborne's house of verandahs in the rainforest, where birds visit all day long; Harry's shack, home to a gecko, a skink, green tree frogs and green ants. These houses, often made of the stuff of the abject, are permeable to the universe from inside and out. Or, perhaps, even more in Nganyinytja's 'front doorstep' the stone hill which marks the threshold of her *ngurra*/country, or Maureen's front gate where she claims her land and her being in the landscape.

And is this a gesture, as Paul Carter (1992b:125) says, in which 'the soul must pass out of the body and pass to the other side, if there is to be conversation'?

Afternoon tea

One afternoon Laura carries a beautifully sculpted orange-sour-cream-cake, shiny golden on bright blue plate, to Sylvia and Lizzie's for afternoon tea. The shining cake is placed in the centre of a table surrounded by four chairs where we sit and chat across the space of the cake. *Take eat this is my body* ... Laura's cake, like countless cakes brought similarly to afternoon teas, becomes the vehicle in a performance of self.

> In what sense is gender an act? As in other ritual social dramas, the action of gender requires a performance that is repeated. This repetition is at once a reenactment and reexperiencing of a set of meanings already socially established; and it is the mundane and ritualised form of their legitimation (Butler 1990: 141).

How does this peformance go when, rather than being 'the mundane and ritualised form of their legitimation', women rewrite the script of this performance?

Laura says she learned to build when she realised *it was just dirt*. I ask her about making mudbricks:

It's like something you've cooked all the time, it changes. At the beginning we tried to do it scientifically. We got soil samples from the quarries and put it in Vacola bottles and shook it with water and the dirt settles out into layers so you can see what proportion of sand and clay there is. But you get to know the dirt, and you relate to the dirt differently. The recipe for the mudbricks depends on that—the ingredients shift slightly, it works like a cake.

Recipe for 5 blocks, 10 x 15 x 5 inches and 50 lbs each
15 shovelfuls of dirt
1 flat shovelful of cement
enough water to mix to a mud pie consistency
Once mixed, fill the mould with the mud and tamp down. You can take the mould off immediately and the brick slumps a bit, giving a lovely tummy bulge. You can add straw to help spread the stress load. In the beginning you do it by the book, scientifically, but in doing it you know it differently, you know it in your body.

~

Marie and May also describe making their houses as a simultaneous act of building and inhabiting.

Marie: Mummy used to make the goody and then she'd say 'You kids paint the walls.' Goody's flour and water, you mix it up. You make a damper, right, but in the goody you put more water than the flour and in the damper you put more flour than the water. Righto, I 'ad Kevin and I 'ad Robert and we was only little then up behind the Showground and Kevin was just startin' kindergarten, and we used to get the charcoal from the fireplace and write the clocks on the wall after we'd goody-ed the wall. We'd always make Mummy give us one square of this paper, we'd go down and buy the white paper, that was our blackboard and we learned each other. Kevin was only little and we'd make 'im say a-b-c and draw the clock and put on the different times and rub it off with an old bit of rag. And we'd be goin' along page by page and then we could sit up and read the news, what was happenin', on the wall. When the walls'd get wet, the goody 'd let go and you'd 'ave no wall. We used to take the walls over to the creek then and re-do it, start over again.

May: Mum, I think, done more of the building than Dad. We used to put all these kerosene tins up, place them on the roof. A lot of them was cut open that we found. There were other people's houses that moved away and we used to go through the scrub and carry the kerosene tins home. We had a ground floor, the walls inside were kerosene tin of course, and they rusted so when it would get round to Christmas time, we used to paper them. The shops' d give us some paper and we'd get the flour and mix it up with water and we'd do all the walls. Then you'd go outside and pray it didn't rain and wash it off (laughing), and read the news. We could sit down on a chair and read the news and everything else.

~

Ferrier (1990: i) argues that colonisation is primarily a spatial conquest and that space is a crucial, though often suppressed factor, in power relations. She discusses cartography and architecture as primary colonising spatial practices of modernity and suggests that postcolonial transformations can be made through alternative models. Her major strategy is a postcolonial reading practice which reads against the grain, certain characters and architectural spaces of Australian literature:

> Miss Hare in her crumbling mansion, Himmelfarb in his semi-nomadic dwelling, Harland with his tent; all seem to accept fragmentation and disintegration. As subjects in culture they are decentred, continually in process. In these works the ideal of a fixed, unified, cultural identity gives way, like the closed house, to allow for change and to acknowledge differences and contradictions within and between cultures (Ferrier 1990: 255).

Ferrier's reading practice produces a set of stories and characters very similar to mine, except that the characters seem to be passive, finding themselves in these situations rather than creating them. Only Harland is a builder, having to begin again from 'the scribbles on a gum'. And while Ferrier's interest is in a postcolonial reading practice, mine is in a postcolonial practice of writing.

I am looking for a different process of production, one that transforms practices of space in the first place, one that, in Marie's words, *makes it good for ourselves to go forward*. In contrast to Ferrier's, my reading of modernity and postmodernity is spatial rather than chronological. I see postmodernity as modernity's margins, always immanent in modernity, rather than two distinct temporal phases. Forms of production such as bricolage can be characteristic of both modernity and postmodernity, the difference being that within modernity (as a philosophical system) meaning lies in the ultimate production of a seamless whole, while in postmodernity, meaning lies in the space in-between, the intertexts or the surfaces coming together.

I think about my writing as a practice of *fabrication, bricolage and assemblage*. It begins with my body, body as centre, a journal of my body in place. Then gathered around the body are all the fragments that contribute to the construction—other voices, transcripts of their stories, conversations, writings of theory, news items and clippings, new and old photographs. Once I have gathered all the bits, I play with them in relation to each other. The game is broken down into smaller and smaller bits until it is all play, no bit can be so threatening as to become too serious. The pieces of text are pasted onto recycled paper, often overlapping with sticky tape and staples, colour coded with highlighters as I find others that disrupt the pattern. The pile of makings grows into a firm wad and then it is ready to go into the computer with all its cut and paste, erase and highlight functions. Once in the computer there is more movement, more play. Like games in Play School, there is no final meaning, only another game.

But it is a very big movement from my writing practice to Ferrier's *episteme of postmodernity*; from Foucault's *little tactics of the habitat* to *the great strategies of geopolitics*,[2] from the daily acts of inhabiting to postmodern theories of space, from the intimate of the maternal space to the world of the symbolic, and I wonder if I am able to make this movement in my writing and how this movement is constituted?

Night settles

At dusk the frogs begin to sing as the first stars appear in the night sky.

I hear on the news that, in South Australia, people are being asked to record each day, for ten minutes, at dusk, in the same place, the sounds of frogs. Frogs, with their highly permeable skin, are so sensitive to the presence of toxins in the water that their singing is a measure of the health of our waterways.

> Immensity is within ourselves. It is attached to a sort of expansion of being that life curbs and caution arrests, but which starts again when we are alone. As soon as we become motionless, we are elsewhere; we are dreaming in a world that is immense (Bachelard 1969: 184).

Bachelard's idea of intimate immensity is a very simple one really. It is that immensity derives from the intimate, that it is an intensity of the intimate and I understand the intimate as the maternal space, no longer a passive being but a positive space of doing, the act of inhabiting. It is a reversal of Irigaray's woman *entombed at the centre of philosophy*; by naming the maternal space, the basis of immensity, of the general, is knowable through the body. In naming the space of the intimate, the origins of the immense become visible.

---

2 A whole history remains to be written of spaces—which would at the same time be the history of powers ... from the great strategies of geo-politics to the little tactics of the habitat (Foucault cited by Spain 1992: 1).

And the spatial intimate is what I have called the function of inhabitation.

I think of Patsy's description of her grandmother's house, built around the hearth, but with holes in the roof through which you can see the stars.

At night I ask Malcolm a little about the stars, and the observatory on the mountain. It is why Laura and Malcolm live here; it makes rhythms in our daily lives, how can I not write about it in this story of inhabitation. Sometimes when the moon is dark Malc works all night observing and when the moon is too bright, around full moon, there is no observing at all. He talks about 'the mountain': 'It snowed up the mountain.' 'The dust storm was amazing up the mountain.' Why is the observatory here on this mountain?' I ask. 'Because it is high,' he says, (up the mountain), 'It is clear [a long way from any major sources of pollution], visibility is good [no competing light sources] and there is a good chill factor.' I ask about the Pleiades, a group of stars I can name, the Seven Sisters from Pine Gap, from Nganyinytja in the desert. 'Can you see them in the sky here?' At the moment they are rising late, but by November they will be visible in the early evening. I ask Malcolm about stories of the Pleiades. He doesn't know any stories he says. I feel that I don't know the right questions to ask, the right language, maybe I'll have to wait until open day at the Observatory.

At the Observatory, on top of 'the mountain', we stand in cold winds in a revolving cage on the biggest telescope, and survey miles of landscape lying at our feet; we hear about the depths of the heavens that can be seen through infra-red light, beyond the field of human vision; we look at photographs of nebulae where stars are born and explosions of supernovae where stars die.

> Only the largest stars can turn into supernovae ... Quite suddenly the star finds it cannot sustain the activity. The inner portion collapses to make a neutron star and a violent wave ripples out through the remainder of the star ... in the process

the ripple turns the material into a mixture of elements such as iron, lead, silver, gold. Virtually all the metals in the building around you, the iron in your bloodstream, and all the precious metals ... were created in supernovae that erupted before the Earth was formed (sign from display at Coonabarabran Observatory).

Perhaps we are made of stardust.

Overwhelmed by the time we get to the Schmidt Telescope, I am glad to see Malcolm standing beside a large coloured photograph of the Pleiades. I ask him about the seven stars. 'There's hundreds of stars,' he says, 'The seven you can see with the naked eye are the brightest and through a telescope each of the seven is a cluster of stars that opens out into hundreds more stars.' He points out the swirling nebulous mass around each of the seven main stars and their blueness. 'Blue is the colour of young stars,' he says, 'and the nebulosity is a swirling mass of energy in which new stars are born.'

But still, at the heart of the matter, in the white domes of knowledge, like the domes at Pine Gap, there is an absence, the I/eye is always focused elsewhere. Where is the intimate of this immense?

Laura and I get to know each other better and better. 'Who is that living in your knitting,' she says, and gently lifts out a small creature that turns out to be a baby grasshopper taking shelter between the squares of my blanket.

> here we discover that immensity in the intimate domain is intensity, an intensity of being, the intensity of being evolving in a vast perspective of intimate intensity. It is the principle of correspondences to receive the immensity of the world, which they transform into intensity of our intimate being. They institute transactions between two kinds of grandeur (Bachelard 1969: 193).

Laura unravels as I knit and we talk, and we look up stories of the Pleiades. There are several stories of European origin of transcendence and (romantic) love but the ones I love best are the Aboriginal stories of corporeality, sexuality and the cycles of the earth:

> *Artunyi*, the Seven Sisters
> At Yandara in NSW, a great serpent called *Akurra* ate up a whole camp of *Yuras*. Among these was a group of women called *Artunyi*.
> Later on a big storm burst over the land, and in the floods that followed, the serpent was drowned. The storm went on for weeks and weeks, and the water rose so much it almost touched the sky.
> The dead *Akurra* who was floating on top of the water, finally burst because his body rotted, and this blew the *Artunyi* into the sky where they have remained ever since (Tunbridge 1988: 15) (my italics).

Another story of creation from the abject body.

> *Artunyi*, the Pleiades, played an important part in *Adnyamathanha* traditional life ... In particular they marked the seasons. They disappear from the sky in winter (around May), and in the evening do not reappear until November. They rise in the early morning in July, however, and this signifies that it is *malkada* time—the time for the ceremony when boys are made *Vardnapas*.
> The appearance of *Artunyi* up over Wayanha (Mount McKinlay) marked the beginning of the frost in *Adata Madapa* (Frost Valley where Nepabunna is situated). The two celestial *Urndakarra* who kept their hair in place with a net would let it down when the *Artunyi* appeared in the east. It reached to the ground thus making the frost ... They then collected the ice on the water and the frost on the ground and ate it and rubbed it all over themselves so that the girls would grow large breasts and the boys would grow long beards (Tunbridge 1988: 16).

A story of inhabiting the stars, bodies to stardust, stardust to bodies.

~

Back in Armidale I talk about my ideas with other men. I want to know what they think, try to find the right words to cross this space of difference. I talk about liminality, and Robert says, 'Yes, I think Australia is still in a liminal space. Australians are becoming more Aboriginal,' he says, 'They are becoming more mobile, less interested in possessions, wanting more to explore themselves as part of this landscape rather than pitted against it.' And I say 'What about our new breed of feral children, who sit in trees in the south-east forest, resist ownership, live in squats and teepees, or even cardboard boxes?'

And Graeme returns from the desert with stories of living with *Tjutupayi*, a very old man; waking at dawn and sitting naked to feel the air moving on skin, where the breezes are coming from that day; and of walking five hundred kilometres from Docker River to *Uluru* and crossing the paths of yogis travelling to *Uluru* to meditate on sunrise at the rock.

Notes towards a practice of love

In writing about houses and what makes a house a home, I begin to think about a practice of love. There is an image that comes to me now from *Dingo Makes Us Human*, of a woman sitting in the landscape among tall grass seed, trees in the background, looking with great attention at a special plant in her hand (Rose 1992). The caption says 'Jessie Wirrpa: taking exquisite notice and care, March 1986'. This is a practice of love. How can we begin to speak that practice?

Elspeth Probyn says:

> I insist on a concept of the self that can articulate the theoretical necessity of care, of love and of passion (1993b: 169).

Practices of the body
Body painting, body dancing (Pintubi women in the desert).

Massage:

> like finding the hills and rivulets—tracing the more subtle connections which give meaning to the landmarks of the body (Carmont 1996b: pers. comm.).

Body in place, place in body.

Placing the body at the centre of theory:

> putting the body at the centre of our notion of subjectivity transforms the way we think about knowledge, about

power, about desire ... and this entails the possibility of forming other kinds of knowledge, other kinds of social interrelations, other forms of ethics, other systems of representation based on different interests, not only those of women, but those of cultural others ... whose bodies are inscribed in different forms and therefore whose subjectivities and intellectual frameworks are different (Grosz 1992b: 5).

Care of the self:

I want to set out ways of embodying a care of the self in speech and in writing (Probyn 1993b: 4).

Writing the body, body writing:

So does certain women's womb writing, which neither separates the body from the mind nor sets the latter against the heart ... but allows each part of the body to become infused with consciousness. Again, bring a new awareness of life into previously forgotten, silenced, or deadened areas of the body (Trinh 1989: 40).

Practices of the moon, dreams, poetry and body cycles (Pine Gap) and the intervention of poetry into theory.

Practices of the breath:

Become aware of your breath
in both the nostrils
natural breath
flows into the nostrils and flows out of them
bring your mind closer to your natural breath
steadily deepen your awareness
about your physical breath
(*Satyananda* meditation).

Inside and outside, self and other are relativized, porous, each time one takes a breath. The air is constantly transgressing boundaries, sustaining life through interconnection. One may have spent yesterday studying the mystics on the unreality of dualism and have this remain as an abstract idea. But in following the breath one begins to embody this truth (Leder 1990: 17).

Stillness, solitude and retreat. Franz Kafka:

> You do not have to leave the room; remain standing at your table and listen; do not even listen, simply wait; do not even wait, be quite still and solitary; the world will offer itself freely to you, to be unmasked, it has no choice, it will roll in ecstasy at your feet (Kafka cited by Devereux: 1994).

Watching the ants.

Practices of self and other

Storytelling:
Patsy, Emily, Nganyinytja, Maureen, Marie, Janet and May, Maisie, Kathy, and Bill all telling their stories.

> The story depends on every one of us to come into being. It needs us all, needs our remembering, understanding, and creating what we have heard together to keep on coming into being (Trinh 1989: 119).

> My great-grandmama told my grandmama the part she lived through that my grandmama didn't live through and my grandmama told my mama what they both lived through and my mama told me what they all lived through and we were supposed to pass it down like that from generation to generation so we'd never forget (Jones cited by Trinh 1989: 122).

Laughing at, and in our stories, and 'making it good for ourselves to go forward' (Marie in Somerville and Dundas et al. 1994).

Opening up a symbolic space of exchange:

> Their value lies, not only in bringing people without a language in common together, but in opening up a space between and around them, a dynamic space that, unlike the proscenium of later nineteenth-century theatre, kept all views open; that, unlike the monocular perspective of the ideal theatre-goer, preserved intervals of difference (Carter 1992a: 179).

and devising new movements between these intervals of difference:

> The problem is not to preserve differences but to devise new movements between them. And, if they are to succeed, the movements will have to be as much poetic as political (Carter 1992b: 22).

Moving to the limit-edge of self, towards the other, with attitude, empathy and imagination:

> In order to realise the self as 'limit-attitude' where we work at the very edges and ends of our selves in order to envision change, we must engage our imaginations more fully (Probyn 1993b: 6).

> a limit-attitude stipulates that we work at the frontiers of ourselves, not looking back on what we know, but rather that we look forward to what we do not know (Probyn 1993b: 140).

Imagining what it might be like to experience, to see the world as the other:

> The ability to imagine is ... a pre-condition for the capacity to articulate rhizomatic lines that touch and connect with the aspirations of others (Probyn 1993b: 148).

and empathy, enabled by imagination, to participate fully in the feelings and ideas of others:

an alternative critical practice that would be set in motion by imagination, sustained through empathy, and that seeks the articulation of new spaces in the movement between 'who is she and who am I'. At an epistemological level, empathetic practices of imagination can be used to foreground the fact that we do indeed piece together knowledges in various ways (Probyn 1993b: 148).

Turner's practice of communitas:
- unmediated relationship between historical, idiosyncratic, concrete individuals
- more in contrast than active opposition to social structure as an alternative more liberated way of being socially human
- tends to be inclusive, has something of a flow quality but often arises spontaneously (Turner 1982: 45–50).

Social drama, carnivale, antistructure.

The related practices of deep remembering (Brooks 1991) and singing our history (Griffin 1994).

Practices of home

Nganyinytja's front doorstep.
Marie's 'paper house' and Monday Tuesday Wednesday routine of housework.

> all really inhabited space bears essence of the notion of home (Bachelard 1969: 5).

> nest chrysalis and garment only constitute one moment of a dwelling place (Bachelard 1969: 66).

> Thus the dream house must possess every virtue. However spacious, it must also be a cottage, a dove-cote, a nest, a chrysalis. Intimacy needs the heart of a nest (Bachelard 1969: 65).

Cleaning sweeping dusting scrubbing:

> A house that shines from the care it receives appears to have been rebuilt from the inside (Bachelard 1969: 68).

The pleasure of the washing on the line:

> The wardrobe is filled with linen
> There are even moonbeams that I can unfold (Rimbaud cited by Bachelard 1969: 66).

Collecting preparing cooking feasting eating food:

> Understanding food philosophically not only legitimates new categories of enquiry (the epistemology of food and the ontology of food for example), but also holds the potential to provide further illumination of traditional philosophical problems (such as the relation between thought and practice or mind and body, and the concept of a person) (Curtin and Heldke 1992: xiii).

Eating bushtucker, eating place: octopus and mussels, turtle eggs, pippis, five corners and fiery hogs.

Displacement and exile; carrying home inside us wherever we go:

> for a number of writers in exile the true home is to be found not in houses but in writing. Home has proven to be here both a place of confinement and an inexhaustible reservoir from which one can expand (Trinh 1993: 7).

Sweeping the graves clean.

Practices of play

Knitting a blanket, transitional objects and play:
A transformative aesthetics based on the transitional object (in-between-self-and-other space):

> The transitional object, a thing of paradox ... opens up a third space, the site of of play (Armstrong 1993: 183).

> The transitional object mediates a life-creating, culture-modifying space which is at once transgressive and communal (Armstrong 1993: 184).

Nevertheless our play is *play*, not simply subversive linguistic play, but the transformation of categories which constitutes a change in the structure of thought itself: it is not only an aspect of knowledge but the prerequisite of political change (Armstrong 1993: 184).

Beach as pleasure and holidays and beach as margin and landscape. Moving to landscape across the space of difference with imagination, empathy and attitude, the same movement between self and other as practised by Probyn.

Modjeska's seeing from the heart (1994) and brooding patterns of thought (1990),
looking at the stars,
naming the intimate in the immense.

Playing with words:

> *lingalonga over lingua*
> *you leave me reader working on the body of my new*
> *lover Trixi oh sorry what was it again Lexi yes Lexi Con*
> *holding her spine in the palm of my hand*
> *ah and ohyes the body of language*
> (Fallon 1989: 32)

Music, women singing, the sea
immersion and connection
movement, unsettlement
sexuality and refiguring female desire
love of landscape.

Practices of body/landscape
Song, dance, story, tracking.
Singing the land,
Dancing the country
Storying place.

Tracking as an alternative spatial practice to mapping:

> *iwara* is the mark of animals, including humans on the ground as they move about their lives; it is also the track that connects one place to another, including roads; and it is the visible material representation of the Dreaming ancestor moving through the country along a songline (Pitjantjatjara Dictionary).

> In the Aboriginal science of tracking, following someone's footsteps means to know the person. To walk exactly in their footsteps means that there is an imitation—not a reproduction, of the whole movement of their bodies (Benterrak et al. 1984: 223).

Space mapped through multiple stories:

> alternative cartographies can *construct* rather than *represent*, space as heterogeneous, open, defined through multiple viewpoints, through many particular localised operations and contacts (through tactile, not just visual means) (Ferrier 1990: 40).

Songlines, mapping by a linking between stories:

> The exchanges, whether of sounds, gestures or objects, are like surveyors' lines connecting horizon points: their thickening network of sightlines, orienting this person to that person, this face to that face, help to map a common space, to characterise its behavioural and symbolic topography (Carter 1992a: 179).

Practices of place:
Painting wildflowers,
growing rainforest from cassowary droppings,
maintaining networks of place knowledge.

Body/place:
Living on the ground
and learning a language of place.

Breathing the air of trees,
walking with 'exquisite notice and care',
listening with the ruffled surface of the body.

Healing the land, as Nganyinytja describes it:

> The two laws need to become one to keep the land. We, the Pitjantjatjara people have always kept our land and looked after it and make it grow ... If people will listen to our way they will understand why we live in the country of our grandparents and why we must have strong land rights (cited by Macken 1993: 23).

References Cited

Anderson, Benedict. 1983. *Imagined Communities: Reflections on the Origin and Spread of Nationalism*. London and New York: Verso.
Andrews, John. 1998. In conversation.
Armstrong, Isobel. 1993. So what's all this about the mother's body?: The aesthetic, gender, and the polis. *Women: A Cultural Review*, 4 (2), 172–187.
Bachelard, Gaston. 1969. *The Poetics of Space*. Boston: Beacon Press.
Ball, Desmond. 1980. *A Suitable Piece of Real Estate: American Installations in Australia*. Sydney: Hale and Iremonger.
Barthes, Roland. 1984. *Camera Lucida: Reflections on Photography*. London: Flamingo.
Barwick, Diane E. 1978. And the lubras are ladies now. In Fay Gale (Ed.), *Woman's Role in Aboriginal Society*. Canberra: Australian Institute of Aboriginal Studies.
Barwick, Linda. 1991. *Group Project on Antikirrinya Women: 1966–1968 expeditions*. Department of Music, University of New England (limited publication).
Bastable, Mary. 1993. The Textual Representation of Life-Story-Telling as Social Theory. Masters Thesis, Department of Social, Cultural and Curriculum Studies, University of New England.
Bell, Diane. 1983. *Daughters of the Dreaming*. Sydney: Allen and Unwin.
Benterrak, Krim and Stephen Muecke, Paddy Roe and others. 1984. *Reading the Country: An Introduction to Nomadology*. Fremantle, W.A.: Fremantle Arts Centre.
The Bible. King James version.
Bilson, Gay. 1994. The Blood of Others. *Sydney Morning Herald, Good Weekend*, May 14.
Blau du Plessis, Rachel. 1990. *The Pink Guitar: Writing as Feminist Practice*. New York: Routledge.
Brock, Peggy (Ed.). 1989. *Women, Rites and Sites: Aboriginal Women's Cultural Knowledge*. Sydney: Allen and Unwin.

Brooks, David. 1991. Literature and Forgetting. *Southerly 3*.
The Bush Brother. 1909. *Journal of the Bush Brother*. A Missionary Journal held at the Library of the Aboriginal & Torres Strait Islander Studies.
Bryce, Suzy. 1992. *Women's Gathering and Hunting in the Pitjantjara* [sic] *Homelands*. Alice Springs: Institute of Aboriginal Development.
Butler, Judith P. 1990. *Gender Trouble: Feminism and the Subversion of Identity*. New York: Routledge.
Cain, Mary J. n.d. Letter to the Governor of New South Wales. Archives of New South Wales.
Cain, Mary J. 1920. Recollections of Coonabarabran. Manuscript. State Library of New South Wales (Mitchell Library).
Carmont, Cathy. 1996a. Tissue Talk. PhD Thesis, Department of Social, Cultural and Communication Studies, University of New England, Armidale.
Carmont, Cathy. Written pers. comm. 1996b.
Carter, Paul. 1990. Ghosts. Editorial. *Age Monthly Review*, May.
Carter, Paul. 1992a. *Living in a New Country: History, Travelling and Language*. London: Faber and Faber.
Carter, Paul. 1992b. *The Sound in-Between: Voice, Space, Performance*. Sydney: University of New South Wales Press and New Endeavour Press, Sydney.
Castenada, Carlos. 1968. *The Teachings of Don Juan: A Yaqui Way of Knowledge*. California: University of California Press.
Chapman, Tracey. Album. 'Across the Tracks'.
Cixous, Hélène. 1981. The Laugh of the Medusa. In Elaine Marks, and Isabelle de Courtivron (Eds.), *New French Feminisms: An Anthology*. New York: Schocken.
Cohen, Patsy and Margaret Somerville. 1990. *Ingelba and the Five Black Matriarchs*. Sydney: Allen and Unwin.
Connors, Emily. n.d. Manuscript.
Couani, Anna and Barbara Brooks. 1983. *The Train and Leaving Queensland*. Glebe, NSW: Sea Cruise Books.
Curtin, Deane W. and Lisa M. Heldke (Eds.). 1992. *Cooking, Eating, Thinking: Transformative Philosophies of Food*. Bloomington: Indiana University Press.
de Lauretis, Teresa. 1994. *The Practice of Love: Lesbian Sexuality and Perverse Desire*. Bloomington: Indiana University Press.
Denzin, Norman K. and Yvonna Lincoln (Eds.). 1994. *Handbook of Qualitative Research*. California: Sage Publications.
Devereux, Lesley. 1994. Unpublished talk at Women and Spirituality Conference, Mullumbimby, June.

Dillard, Annie. 1974. *Pilgrim at Tinkers Creek*. New York: Harper and Row.
Diprose, Rosalyn and Robyn Ferrell. 1991. *Cartographies: Poststructuralism and the Mapping of Bodies and Spaces*. Sydney: Allen and Unwin.
Douglas, Mary. 1966. *Purity and Danger: An Analysis of the Concepts of Pollution and Taboo*. London: Ark Paperbacks.
Downing, C. 1989. *Journey through Menopause: A Personal Rite of Passage*. New York: Crossroad.
Drewe, Robert (Ed.). 1993. *The Picador Book of the Beach*. Sydney: Pan Macmillan.
Ellis, Catherine. 1985. *Aboriginal Music: Education for Living. Crosscultural Experiences from South Australia*. St Lucia: University of Queensland Press.
Ellis, Catherine and Linda Barwick. 1989. Antikirinja Women's Song Knowledge 1963–72: Its significance in Antikirinja culture. In Peggy Brock (Ed.), *Women, Rites and Sites: Aboriginal Women's Cultural Knowledge*. Sydney: Allen and Unwin.
Fallon, Mary. 1989. *Working Hot*. Melbourne: Sybilla Co-operative Press and Publications.
Farmer, Beverley. 1990. *A Body of Water: A Year's Notebook*. St. Lucia: University of Queensland Press.
Fennell, M. and Alex Grey. 1974. *Nucoorilma*. Sydney: Department of Adult Education, University of Sydney.
Ferrier, Elizabeth. 1990. Mapping the Space of the Other: Transformations of Space in Postcolonial Fiction and Postmodern Theory. PhD Thesis, English Department, University of Queensland.
Fusco, Coco. Sankofa and Black Audio Film Collective. In Russell Ferguson, William Olander, Maria Tucker and Karen Fiss (Eds.), *Discourses: Conversations in Postmodern Art and Culture*. New York: New Museum of Contemporary Art.
Gilmore, Mary. 1931. *The Rue Tree: Poems by Mary Gilmore*. Melbourne: Robertson and Mullens.
Greenler, Robert. 1980. *Rainbows, Halos and Glories*. Cambridge: Cambridge University Press.
Griffin, Susan. 1978. *Woman and Nature: The Roaring Inside Her*. New York: Harper and Row.
Griffin, Susan. 1982. *Made From this Earth. Selections from Her Writing, 1967–1982*. London: Women's Press.
Griffin, Susan. 1994. *A Chorus of Stones: The Private Life of War*. London: Women's Press.
Grosz, Elizabeth A. 1989. *Sexual Subversions: Three French Feminists*. Sydney: Allen and Unwin.

Grosz, Elizabeth A. 1990. *Jacques Lacan, A Feminist Introduction*. Sydney: Allen and Unwin.

Grosz, Elizabeth A. 1992a. Merleau Ponty and Irigaray in the Flesh. Transcript of paper presented at the University of New England.

Grosz, Elizabeth A. 1992b. Refiguring Lesbian Desire. Trancript of paper delivered at Australian Women's Studies Conference, Brisbane.

Grosz, Elizabeth A. 1994. *Volatile Bodies: Towards a Corporeal Feminism*. St Leonards, NSW: Allen and Unwin.

Grosz, Elizabeth A. 1995. *Space, Time and Perversion: The Politics of Bodies*. St Leonards, NSW: Allen and Unwin.

Hall, Nor. 1980. *The Moon and the Virgin: Reflections on the Archetypal Feminine*. New York: Harper and Row.

Haug, Frigga (Ed.). 1987. *Female Sexualization: A Collective Work of Memory*. London: Verso.

Heilbrun, Carolyn. 1988. *Writing a Woman's Life*. London: Women's Press.

Holloway, Barbara. 1993. There or No Where: A Study of the Use of Race and Gender Imagery in the Topography of Possession, Australian Poetry 1800–1938. PhD Thesis, University of New England, Armidale.

hooks, bell. 1991. Narratives of Struggle. In P. Mariani (Ed.), *Critical Fictions: The Politics of Imaginative Writing*. Seattle: Bay Press.

Hulme, Keri. 1993. Unnamed Islands in the Unknown Sea. In Robert Drewe (Ed.), *The Picador Book of the Beach*. Sydney: Pan Macmillan.

Irigaray, Luce. 1981. This Sex Which is Not One. In Elaine Marks and Isabelle de Courtivron (Eds.), *New French Feminisms: An Anthology*. New York: Schocken Books.

Irigaray, Luce. 1991. *Marine Lover: of Friedrich Nietzsche*. New York: Columbia University Press.

James, Diana. Written pers. comm. 1984.

Jones, Rhys. 1991. Landscapes of the Mind: Aboriginal perceptions of the natural world. In Derek John Mulvaney (Ed.), *The Humanities and the Australian Environment*. Canberra: Highland Press.

Kirby, Vicki. 1991a. Transcript of paper presented at the Body Politics Conference, University of Sydney.

Kirby, Vicki. 1991b. Corporeal Habits: Addressing Essentialism Differently. *Hypatia* 6, Fall.

Leder, Drew. 1990. *The Absent Body*. Chicago: University of Chicago Press.

Lloyd, Patricia J. 1988. Politics at Pine Gap: Women, Aborigines and Peace. Bachelor of Letters thesis. Deakin University.

Lorde, Audre. 1996. *The Audré Lorde Compendium: Essays, Speeches and Journals*. London: HarperCollins*Publishers* (1996 edition).
McAllister, Pam (Ed.). 1982. *Reweaving the Web of Life: Feminism and Nonviolence*. Philadelphia, PA: New Society Publishers.
McArthur, Kathleen. (n.d.). *Living on the Coast*. Kangaroo Press, Australia.
McConaghy, Cathryn. 1999. Aboriginal Education Unit Notes, University of New England.
McDonald, Christie. 1985. *The Ear of the Other: Otobiography, Transference, Translation: Texts and Discussions with Jacques Derrida*. New York: Schocken Books.
Macken, Deirdre. 1993. Aboriginal Reconciliation: Time to Get Serious. *Sydney Morning Herald, Good Weekend*, August 21.
Marcus, George E. 1994. What Comes (Just) After 'Post'? The Case of Ethnography. In Norman K. Denzin and Yvonna S. Lincoln (Eds.), *Handbook of Qualitative Research*. California: Sage Publications.
Marks, Elaine and Isobelle de Courtivron. 1981. *New French Feminisms: An Anthology*. New York: Schocken.
Meyerhoff, Barbara. 1986. Life Not Death in Venice: Its Second Life. In Victor W. Turner. and Edward M. Bruner (Eds.), *The Anthropology of Experience*. Urbana: University of Illinois Press.
Modjeska, Drusilla. 1990. *Poppy*. South Yarra, Vic: McPhee Gribble.
Modjeska, Drusilla. 1994. *The Orchard*. Sydney: Macmillan.
Morrison, Toni. 1987. *Sula*. New York: New American Library.
National Parks and Wildlife Service. 1976. *Survey of Sites of Significance to the Aboriginal People of NSW*. Survey Ref GM/I/S—10/99.
Neville, Fionnuala. 1999. In conversation.
New Imperial Reference Dictionary. London: George Newnes.
Noonan, Anne. 1996. Psyche and Environment. Papers presented at a Conference on Sense of Place: Depth Perspectives on Australian Landscapes and Environmental Values. University of Western Sydney, Hawkesbury.
Ong, Walter J. 1982. *Orality and Literacy: The Technologising of the Word*. London; New York: Methuen.
Oxley, John. 1820. *Journals of Two Expeditions into the Interior of New South Wales*. London: J. Murray.
Payne, Helen. 1988. Singing a Sister's Sites: Women's Land Rites in the Australian Musgrave Ranges. PhD thesis, Department of Music, University of Queensland.
Payne, Helen. 1989. Rites for sites or sites for rites? The dynamics of women's cultural life in the Musgraves. In Peggy Brock (Ed.), *Women, Rites and Sites: Aboriginal Women's Cultural Knowledge*. Sydney: Allen and Unwin.

Plowman, Kristine. Written pers. comm. 1994.

Probyn, Elspeth. 1993a. True Voices and Real People: The Problem of the Autobiographical in Cultural Studies. In Valda Blundell et al. (Eds.), *Relocating Cultural Studies*. London; New York: Routledge.

Probyn, Elspeth. 1993b. *Sexing the Self: Gendered Positions in Cultural Studies*. London; New York: Routledge.

Richardson, Laurel. 1992. The Consequences of Poetic Representation: Writing the other, Rewriting the Self. In Cath Ellis and Michael G. Flaherty (Eds.), *Investigating Subjectivity: Research on Lived Experience*. Newbury Park: Sage Publications.

Reid, Catherine. 1982. Reweaving the Web of Life. In Pam McAllister (Ed.), *Reweaving the Web of Life: Feminism and Nonviolence*. Philadelphia, PA: New Society Publishers.

Roberts, Michèle. 1991. *In the Red Kitchen*. London: Minerva.

Rose, Deborah Bird. 1992. *Dingo Makes Us Human: Life and Land in an Aboriginal Culture*. Cambridge: University of Cambridge Press.

Rushdie, Salman. 1985 (English trans). Introduction. Gunter Grass. *On Writing and Politics 1967–1983*. San Diego: Harcourt Brace Jovanovich.

Sacks, Oliver. 1986. *The Man Who Mistook His Wife for a Hat*. London: Picador, Pan Books.

St Vincent Welch, Sarah. 1985. From a Journal—1983-4. In *Minute to Midnight: New Writing for Peace and Disarmament*. Sydney: Red Spark Books.

Saraswati, Swami. 1973. *Asana, Pranayama, Mudra Bandha*. Bihar, India: Bihar School of Yoga.

Silkwood, Karen. 1983. Album. Director Mike Nichols. Distributed by Roadshow Entertainment.

Schechner, Richard. 1986. Preface. Victor Turner's Last Adventure. In Victor Turner *The Anthropology of Performance*. New York: PAJ Publications.

Smith, Sidonie. 1987. *A Poetics of Women's Autobiography: Marginality and the Fictions of Self-Representation*. Bloomington: Indiana University Press.

Somerville, Margaret and Marie Dundas, with May Mead, Janet Robinson and Maureen Sulter. 1994. *The Sun Dancin': People and Place in Coonabarabran*. Canberra: Aboriginal Studies Press.

Somerville, Margaret with Florrie Munro and Emily Connors. 1995. In Search of the Queen. In Linda Barwick, Alan Marett and Guy Tunstill (Eds.), *The Essence of Singing and the Substance of Song: Recent Responses to the Aboriginal Performing Arts and Other Essays in Honour of Catherine Ellis*. Oceania Monograph No. 46. Sydney.

Spain, Daphne. 1992. *Gendered Spaces*. Chapel Hill; London: University of North Carolina.
Trinh T. Minh-ha. 1989. *Woman, Native, Other: Writing Postcoloniality and Feminism*. Bloomington: Indiana University Press.
Trinh T. Minh-ha. 1991. *When the Moon Waxes Red: Representation, Gender and Cultural Politics*. New York; London: Routledge
Trinh T. Minh-ha. 1992. *Framer Framed*. New York; London: Routledge.
Trinh T. Minh-ha. 1993. *Gender and the Poetry of Desire*. Keynote Address, Forces of Desire Conference, Canberra.
Tunbridge, Dorothy. 1988. *Flinders Ranges Dreaming*. Canberra: Aboriginal Studies Press.
Turner, Victor. 1982. *From Ritual to Theatre: The Human Seriousness of Play*. New York: Performing Arts Journal Publications.
Turner, Victor. 1986a. Dewey, Dilthey, and Drama: An Essay in the Anthropology of Experience. In Victor Turner and Edward M. Bruner (Eds.), *The Anthropology of Experience*. Urbana: University of Illinois Press.
Turner, Victor. 1986b. *The Anthropology of Performance*. New York: PAJ Publications.
Turner, Victor. 1990. Are there universals of performance in myth, ritual and drama? In Richard Schechner and Willa Appel (Eds.), *By Means of Performance: Intercultural Studies of Theatre and Ritual*. Cambridge; New York: Cambridge University Press.
Turner, Victor and Edward M. Bruner (Eds.). 1986. *The Anthropology of Experience*. Urbana: University of Illinois Press.
Wafer, Petronella (Ed.). 1990. *History and the Land: A Story Told by Yami Lester*. I.A.D. (Institute of Aboriginal Development) Alice Springs.
Walker, Michèle. 1993. Infidelity, Silence and Reason: the debate between Luce Irigaray and Michelle Le Doeff. Unpublished paper, Women and Philosophy Conference, University of Queensland.
Wilcox, Helen, Keith McWatters, Anne Thompson and Linda R. Williams. 1990. *The Body and the Text: Hélène Cixous, Reading and Teaching*. New York: Harvester Wheatsheaf.
Winton, Tim. 1993. *Land's Edge*. Chippendale, NSW: Pan Macmillan in association with Plantagenet Press.
Woolf, Virginia. 1945. *A Room of One's Own*. Harmondsworth: Penguin Books.
Woolf, Virginia. 1976. *Moments of Being: Unpublished Autobiographical Writings* [by] Virginia Woolf, edited by Jeanne Schulkind. London: Chatto and Windus for Sussex University Press.

Index

abjection 151–2, 171, 191–2
Aboriginal art 73–4
Aboriginal culture 8
Aboriginal oral stories 5, 8–9
   and academic voice 9
   representation as written text 10–11
   see also under specific oral historians, eg. Marie's stories
Aboriginal women, at Pine Gap 38–9, 40–1
academic accommodation, author's 85–6
anangu see Pitjantjatjara
Anderson, Benedict 5
Angatja artworks 74
Angatja landscape 44–8
   camp site 50, 52, 67
   desert life 50–1, 52–3
   singing and dancing 55–8, 62–3
Antikirrinya women's songs 37
antistructure 22
Araluen Centre 73
Aranda women 41
archaeological accounts, Warrumbungles 111
archaeological spaces, caves as 188–9
Armstrong, Isobel 223–4
Arrente women 23, 24
Artunyi (Seven Sisters) 216
Australian landscape, sense of belonging in 128–9

Bachelard, Gaston 182, 186, 191, 222

   function of inhabiting 196, 198, 199
   intimate immensity 213–14
Barthes, Roland 29, 30
beach
   as sense of space 174, 193
   author's anticipation of going to 151
   fossicking along the 171–2
   tyre marks on 171, 173
   see also Mullaway Beach; sea
beach house, at Caloundra 157–9
Bilson, Gay 204
blanket squares 149–50, 151–2, 153, 178–9, 215–16
bodily experience 79–81
body
   and abject 151–2
   in space 5
   practices of the 218–20
body/landscape, practices of 224–6
body/place connection 12, 13–14
Brooks, David 28, 222
Bunawanjin Mountain 95, 96, 98
bunya pine 203
Burrabeedee 10, 101, 108–10, 115–16, 144
   author's role in recording stories 130–1
   cemetery 116–18
   Charlotte's story 119, 136–7
   Christmas feast 203–4
   daily household routine 197
   dancing 136

Ethel's story 122–3
Janet's story 125–6, 134
Marie's story 121–3, 131–2, 134–5, 197, 210–11
Maureen's story 119–21
May's story 123–4, 134, 211
missionary presence 118–19, 126
sun dancing ritual 126–7

Cain, Mary Jane 108, 109, 110, 111, 123, 124, 126, 138
  experiences of Coonabarabran 113–14
  grave 117
  land rights claim 1893 115
  Queen of Burrabeedee 115–16, 122
  story of 112–13, 140
Caloundra 157–9
Carmont, Cathy 13–14
carnivale 22
Carter, Paul 17, 78, 84, 89, 90, 92–3, 94, 182, 207, 209, 221, 225
cassowary project 164–5, 167, 168–9, 176, 207
cassowary women, Mission Beach 146, 154, 167–71, 173
caves 186–9
ceremonies 203
Charlotte's stories 119, 136–7, 143
Cixous, Hélène 141, 148, 154, 174–5
Cohen, Patsy 8–9, 129, 143, 144, 183, 184
  grandmother's hearth 189–91, 214
collective of women, images 30–1
colonisation 5, 211–12
Connors, Clariss 96
contact history 84, 92
cooking 200, 201–2, 223
Coonabarabran 108, 114, 116, 118, 124, 127, 128, 144
  author's presence in the landscape 130–1
  book contract signing 135–6
  caves 187
  Endowment Day 134

Coonabarabran women 10, 15
  see also Charlotte; Ethel; Janet; Marie; Maureen; May
coral spawning 161–2, 193, 194
Cossington Smith, Grace 208
cottage *see* Laura's cottage
creation stories 166–7, 216

dancing
  at Burrabeedee 136–7
  by men 56
  by women 7–8, 23–4, 42, 54–5, 56–7, 63
  issue of bare breasts 56–7
Davy, Wee 6
desert environment 50–1, 52–3
  camp site 50, 52, 67, 186
  dancing 56
  singing 55
Desert Tracks (tours) 49, 50, 60
dispossession 16, 131

Ellis, Catherine 37, 94
Emily 76–106, 197–8
  author transcribes stories 103–4
  author's meetings with 81, 82, 105, 106
  autobiography 105, 106
  death of 105–6
  metaphor of Queen's visibility/invisibility 101–2
  Mooki property and mountain 78–9, 81, 104
  space/image of the Queen 94, 98, 100, 103, 104
  stories of the Queen 76, 81, 82, 84–5, 93, 98
  theoretical interpretation 76–8, 79, 80–1, 84, 90–3
  visit with author to Queen's burial site 86, 87–9, 91, 94–5
Emily (Burrabeedee) 139
Ernabella Mission School 58, 59, 60, 62
Ethel's stories 122–3, 137–8

farmers 4–5

father-in-law, memories of war
  27–8
female desire, refiguring 173–4
Ferrier, Elizabeth 5, 211–12, 213,
  225
food and cooking images 199–202,
  203, 204–5, 209–10, 223
Forky Mountain 112, 113, 115, 116,
  130, 144
  walk up 138–9, 140, 141
Fuller, Mrs 143–4
Fusco, Coco 22

Gallipoli, memories of 27–8
graffiti 31, 32
grasslands 2, 4–5, 203
Griffin, Jinnie and Eugene 114
Griffin, Susan 80–1, 222
Grosz, Liz 129, 148, 151, 152, 157,
  172, 173, 174, 205, 219

hearth, as mother 189–91, 214
Hinton, Kathy 141–4
Holloway, Barbara 191, 192
houses
  and self-portrait images 209
  as inside spaces 180
  as maternal 191
  building of 198–9, 210–11
  departure from 194–5
  hearth-as-mother 189–91
  household routine 197, 222
  images of 190–2
  making from the inside out 198
  practices of 222–3
  Trinity Beach 193–4
  see also Laura's cottage
Hulme, Keri 172
Hunt, Alison 41, 42
hybrid place/space 4, 10

identity, and place 8–9
Ilyatjari 57, 58
images, expressing 13–14, 15, 16
indigenous loss, guilt for 6
Ingelba 81, 82
Ingelba 8–9, 90, 97, 98, 112, 125, 129,
  143, 144, 183, 189–90, 197
inhabiting, function of 189–90, 196,
  198, 199, 207, 210–11, 214
intimate immensity 213–14
Irigaray, Luce 148, 149, 150, 157,
  191, 192, 205

James, Diana 41, 49, 57, 64, 73, 74
Janet's stories 125, 126, 134
journal writing 14, 15–16

Kafka, Franz 220
Kawambarai Cave 187
Kelly, Maisie 5
Kirby, Vicki 193
Kris 154, 177, 186
  cassowary story 164–5
  recounts sea stories 156, 160
Kristeva 148, 151, 152, 191
Kungka Kutjara (painting text) 75

landscape 2–4
  and Aboriginal stories 5, 8–9, 16
  and images 15
  spatial practices through
    photographs 29–30
Laura 151, 152, 183, 198, 200,
  209–10
Laura's cottage
  afternoon tea 209–10
  and act of inhabiting 189–90
  building of 198–9, 210
  dawn at the 183
  food and cooking images
    199–202, 203
  garden 202
  midafternoon 205–6
  midday 199–201
  morning rising 189–92
  morning walk 195–7
  night settles 213–16
  retreat to 206
  travel to 185
Laurel's stories 139–40, 201
Leder, Drew 129, 220
lilli pillis 163
liminal phase 80, 92

Index

liminal space 20, 217
 and performance 16–17
 of the Queen (Sullivan) 94–5, 98
liminality 31, 79, 80, 91, 95, 174
love of place 4
Lovelock, Bill 11–12

McArthur, Kathleen 157–9, 177
Macken, Deirdre 46, 64, 226
Madge 188–9
Maisie 202
Malcolm 214, 215
mapping 225
Maralinga 38, 41
Marie's stories 121–3, 131–2, 134–5, 137, 197, 210–11, 220, 222
massage talk 13–14
Maureen's stories 111, 119–21
Mavis 202
May's stories 123–4, 127, 138, 211
men dancing 56
men's violence, protest against 28
mind/body split 12–13
Mission Beach 193, 206
 cassowary project 164–5, 167, 168–9, 176
 cassowary women 146, 154, 167–71
 journal 160–1, 163–5, 175–7
 name origins 155
Modjeska, Drusilla 26, 205–6, 207–8, 224
Mooki 76, 78, 81, 104
Mt Duval landscape 2–4
Mullaway Beach 146
 first day 154–6
 second day 159–60
 third day 162
 fourth day 165–6
 fifth day 171–4
 sixth day 174–5
 seventh day 178
Munro, Florrie 95–6

Nganyinytja 41, 90–1, 98, 226
 and author 53, 58
 Angatja landscape 44–8
 attitude to white people 62
 Christian faith 61
 Desert Tracks tours 49–50, 63–4
 on bare-breasted dancing 57
 on women's business 58
 rights to *wati ngintaka* dreaming 67–72, 166–7
 singing and dancing 63
 taped stories 49, 51, 58, 65–7
 use of interpreter/translator 57, 58–9, 60–2
 vision about land 64–5
ngintaka song text 67–72, 73
 creation of the landscape 166–7
noeme of photography 29–30
Noonan, Anne 7

octopus 160
oysters 163

Papunya 7, 40, 73
parents, images/memories of 26, 27
Patsy *see* Cohen, Patsy
Payne, Helen 72, 94
performance
 and act of gender 209–10
 and liminal space 16–17
 paradox of 207
photographs
 and images of landscape 29–30
 and metaphysics of presence 30–1
 Pine Gap 22–3, 28–30, 31–5
Pine Gap, as Arrente and Pitjantjatjara country 24
Pine Gap Women's Peace Camp 16, 17, 18–43
 Aboriginal women at 38–9, 40–3
 actions at 33–5
 author's protest against men's violence 28
 collective of women 30–1
 domestic images 32–3
 memories and feelings 20–1, 22–3
 metaphysics of presence 30–1

moon-talk ideas and images 24–8
photographs 22–3, 28–30, 31–5
protest against Defence facilities 20
storytelling 39–40
theory 18–20
transforming signs 31–2
white women–Aboriginal women conflict 41–3
women dancing 23, 42
Pintubi women 7–8
Pitjantjatjara Lands 44–8
Pitjantjatjara men 56, 57
Pitjantjatjara women 23, 24, 41
  dancing 56–7
place
  and identity 8–9
  relationship to 11
  stories of 4, 8, 17
play, practices of 223–4
Pleiades stories 214, 215–16
popular landscape photographs, and confirmation of presence 29, 30
postcolonial transformations 5, 211–12
practices
  of body/landscape 224–6
  of home 222–3
  of play 223–4
  of self and others 220–2
  of the body 218–20
presence, metaphysics of 29, 30–1
Probyn, Elspeth 46, 50, 51, 54, 93, 94, 172, 218, 221–2

queen
  space of the 99–100
  see also Cain, Mary Jane; Sullivan, Maryanne

radiation poisoning 36–7, 38
reconciliation 64, 226
remembrance 6
Roberts, Michèle 200
Roe Creek 41
women dancing 23–4
Ron (pastor/translator) 56, 57, 58, 59–62
Rose, Deborah Bird 6

Sacks, Oliver 132
Schechner, Richard 91, 92
Schmidt Observatory 214, 215
sea
  as sense of place 147, 148–9, 154, 193
  evolution from the 156–7
  see also beach; Mullaway Beach
sea stories 156–7, 160, 161–2, 163, 172–3
seeing from the heart 206, 208
self, practices of 220–2
self-portraits, interiors as 208–9
Silkwood, Karen 22, 23, 36
social dramas 91, 92
  liminal space of 17, 20
Somerville, Jessie 35–7, 154, 178–9
Somerville, Margaret
  academic accommodation 85–6
  at Warrawong 79, 83
  naturopathic remedy 51, 53–4
  new home 86–7, 88
  perception of Royal Family 99–100
  sickness 48–9
  understands Pitjantjatjara 65–6
songs and singing 37, 54–5, 63
  ngintaka song text 67–72
stars 214–15, 216
Stoney Creek Mission 84, 96–7, 103
Sullivan, Maryanne ('Queen of the tribe') 81, 82, 84–5, 94, 98–9, 112
  burial site 94–5, 106
  genealogy 97–8
  liminal space/image of 98–102, 104
  stories about 95–7
  visit to burial site 86, 87–9
  see also Emily
The Sun Dancin' 10, 15–16, 90, 108–45
  and author's sense of belonging 128–9

archaeological spaces 188–9
difficulties over orality, language and representation 142–4
household routine 197
story origins 126–7

telescopes 214–15
Thorsborne, Margaret 175–7
Torres Strait pigeons 178
tracking 225
Trinh Minh-ha 54–5, 78, 82, 87, 88, 90, 91, 94, 110, 111, 112, 184, 185, 199, 208, 219, 220, 223
Trinity Beach 193–4
truffle oil steak 203
Turner, Victor 16–17, 20, 22, 31, 79, 91–2, 94, 222
Two Women Dreaming 44–75
Two-Women dreaming dance 24, 39, 40, 43, 61

Violet 138–9, 201
visibility/invisibility paradox 101–2, 206–7

Walker, Michèle 191
war, memories of 26–8
Warrawong (author's property) 79, 83
clearance sale 88
sense of loss over 89–90
Warrumbungles 108, 110–11, 185
sandstone caves 186–7
washing, hanging out 196–7, 198, 223
wati ngintaka dreaming story 67–72, 166–7
Wildlife Preservation Society 177, 206
woman
as home 191
as mother 196
images of 192–3
women dancing 7–8, 23–4, 42, 54–5, 56
women's-only spaces 39
Woolf, Virginia 184–5, 196
Wright, Judith 177
writing 14–15, 26
as fabrication, bricolage and assemblage 212–13
at the cottage 183
woman's needs for 184–5
see also journal writing
written text, as representation of oral stories 10–11

If you would like to know more about Spinifex
Press, write for a free catalogue or visit
our home page.

SPINIFEX PRESS
PO Box 212, North Melbourne,
Victoria 3051, Australia
http://www.spinifexpress.com.au/~women